ELEMENTS OF POINT SET TOPOLOGY

JOHN D. BAUM

DOVER PUBLICATIONS, INC.
New York

Copyright © 1964 by Marian Baum.
All rights reserved.

This Dover edition, first published in 1991, is an unabridged, unaltered republication of the work first published by Prentice-Hall, Inc., Englewood Cliffs, N.J., 1964.

Library of Congress Cataloging-in-Publication Data

Baum, John D.
    Elements of point set topology / by John D. Baum.—Dover ed.
       p.   cm.
    Reprint. Originally published: Englewood Cliffs, N.J. : Prentice-Hall, c 1964.
    Includes bibliographical references.
    ISBN 0-486-66826-6 (pbk.)
    1. Topology. I. Title.
QA611.B29   1991
514—dc20   91-15003
                                                                CIP

Manufactured in the United States by Courier Corporation
66826604
www.doverpublications.com

To My Mother

PREFACE

In writing this book I have kept in mind the undergraduate at the small college. The importance of introducing such a student to modern algebra early in his college career has generally been emphasized. There seems to be no similar emphasis, however, on the necessity of introducing topology into the undergraduate curriculum. Feeling, as I do, that functional analysis provides one of the major syntheses in mathematics, and knowing the fundamental importance of both algebra and topology as bases for analysis, I have tried to provide an introductory text which in spirit parallels many of the present texts in modern algebra. I am hopeful that in the not too distant future a one-semester course in topology will be available to most undergraduate mathematics majors, so that a course in modern analysis can also be offered to the undergraduate student.

The spirit of this book is geometric and axiomatic. The reason for emphasizing the geometric approach is that I believe the student already has had substantial experience with geometry. Hence his access to topology will be easiest if he can make use of his geometric intuition. The reason for the axiomatic approach is twofold: first, it parallels the student's experience in modern algebra; second, it keeps this book in harmony with the current trends in mathematics.

Besides the principal aim enunciated above, I have had two secondary aims. Since point set topology in an introductory course serves as a foundation not only for analysis but also for further work in point set topology and in algebraic topology, I have tried to include topics which will serve students who have interests other than analysis. The main sequence of topics is, of course, the set of those which are of importance in analysis. There are, however, digressions from this main sequence into topics which are of lesser importance in analysis, but which will be of interest to students who wish to pursue topology, either point set or algebraic.

In another sense the intended audience has determined the form of the book. The proofs at the beginning are presented in great detail. They will appear overly detailed to the professional mathematician. However, it has been my experience that the student who comes to axiomatic mathematics for the first time through a course of this nature requires a bit of help in the early stages. In the later chapters the student will be more dependent on his own resources because the proofs no longer contain quite so much detail. What has been said about proofs applies equally to problems. At the beginning of the book they are simple, sometimes trivial; whereas in the last chapter particularly, they require considerably more from the student. There has been no effort made to supply original sources for the notions presented in the text. Instead the references have been supplied to suggest further reading to the interested student. In a number of these reference works a detailed list of original sources can be found, so it seemed redundant to include them here.

Having used preliminary forms of the text with students at Oberlin College, I find that the material in the text can be covered without strain in one fifteen-week semester. If it appears that this is too heavy a schedule, some sections may be omitted without disturbing the continuity of the text: Section 8 of Chapter 1, Sections 3 and 4 of Chapter 3, Sections 4 and 5 of Chapter 4, and Section 3 of Chapter 5 may be omitted if a shorter course is necessary. The problems contain many of the essential examples necessary to illustrate the theory, and it is urged that most if not all of the exercises be assigned. The few exercises referred to as "term papers" are suggested as subjects for further investigation by the student and as teaching aids for the instructor who prefers to assign a term paper rather than a final examination of the traditional sort.

Theorems, definitions, lemmas, and corollaries are numbered consecutively in each chapter. Thus the tenth object of this sort in Chapter 3 is numbered 3.10. Thereafter the number 3.10 is used to refer to this object. Exercises are numbered consecutively in each chapter beginning with 1; for example, the twelfth exercise in Chapter 2 is simply numbered 2.12 and is thereafter referred to as Ex. 2.12. The end of the proof of a theorem is indicated by the symbol "∎," rather than by a statement to the effect that the proof is completed.

I acknowledge with gratitude valuable suggestions received from Professors R. H. Bing and M. L. Curtis. I wish to thank also Ruth Edwards and Elizabeth Carter for help in typing the manuscript and a group of students too numerous to mention individually for finding many a typographical error in early versions of the book. In spite of all this, the ultimate responsibility for any errors lies with me.

JOHN D. BAUM

CONTENTS

CHAPTER 0 **PRELIMINARIES, 1**

1. Introduction, 1
2. Sets, 1
3. The Algebra of Sets, 3
4. Euler-Venn Diagrams, 7
5. Relations, 9
6. Infinite Sets, 11
7. Miscellaneous Assumptions Regarding the Real Numbers, 16

CHAPTER 1 **TOPOLOGICAL SPACES—BASIC DEFINITIONS AND THEOREMS, 19**

1. Neighborhood Systems and Topologies, 19
2. Open Sets in a Topological Space, 23
3. Limit Points and the Derived Set, 27
4. The Closure of a Set, 28
5. Closed Sets, 31
6. Subspaces, 36
7. Limits of Sequences; Hausdorff Spaces, 38
8. Comparison of Topologies, 42
9. Bases, Countability Axioms, Separability, 43
10. Sub-bases, Product Spaces, 50

CHAPTER 2 **CONTINUOUS FUNCTIONS (MAPPINGS) AND HOMEOMORPHISMS, 56**

1. Functions, 56
2. Continuous Functions (Mappings), 58
3. Homeomorphisms, 60
4. Product Spaces, 65

CHAPTER 3 **VARIOUS SPECIAL TYPES OF TOPOLOGICAL SPACES (VARIETIES OF COMPACTNESS), 69**

1. Compact Spaces, 69
2. Separation Axioms, 79
3. Countable Compactness, 89
4. Local Compactness, 92

CHAPTER 4 **FURTHER SPECIAL TYPES OF TOPOLOGICAL SPACES (MOSTLY VARIETIES OF CONNECTEDNESS), 98**

1. Introduction, 98
2. Connected Spaces, 99
3. Components, 105
4. Local Connectedness, 107
5. Arcwise Connectedness, 109

CHAPTER 5 **METRIC SPACES, 115**

1. Definitions, 115
2. Some Properties of Metric Spaces, 120
3. Metrization Theorems, 125
4. Complete Metric Spaces, 132
5. Category Theorems, 136

REFERENCES, 145

INDEX, 147

CHAPTER 0

PRELIMINARIES

§ 1 Introduction

Most mathematical fields these days begin with a collection of undefined objects and a set of axioms which govern the behavior of the objects. There are many advantages to such an approach to a mathematical science. Perhaps the greatest of these is that any mathematical system we encounter that obeys the axioms of a particular mathematical field will also obey all the theorems that are true in that field. Since any axiomatic development begins with a set of undefined objects, it is essential that before we start on the axiomatic development itself we study some set theory.

§ 2 Sets

Set theory itself, being a field of mathematics, has an axiomatic development, but we shall not take that view here, but rather shall develop the theory by relying to a considerable extent on our intuition. Following Cantor we take the word "set" to mean "any collection into a whole, M, of definite and separate objects, m, of our intuition or our thought." The objects which are collected to form the set are called elements or members of the set in question; the whole set, however, being thought of as a single entity. Generally we shall designate sets by capital letters, either Roman, A, B, etc., or script $\mathcal{A}$, $\mathcal{B}$, etc., or German, $\mathfrak{A}$, $\mathfrak{B}$, etc. The elements of the sets will generally be designated by lower-case Roman letters.

We have two ways to indicate what the members of a set are. We can, should the set be a small one, simply list the elements of the set. We do this in a consistent fashion, by writing the list in curly brackets, or braces, e.g., $A = \{1, 2, 3\}$, separating each element from the next by a comma. If the set happens to be large this method is cumbersome, and we prefer then to designate the set by giving some common property that all the elements of the set enjoy, e.g., $A = \{x \mid x$ is a positive integer less than $4\}$. Again our notation is consistent; we use braces, indicate a generic element of the set (in this case, x) by some letter followed by a vertical line, and then write the common property which the generic element as well as all the elements of the set enjoy.

We indicate that a particular element is a member of a set by writing, for example, $2 \in A$. We read this as "two is an element of the set A," or "two is a member of A," or simply "two belongs to A." Should an element not belong to a set, we write for example, $4 \notin A$, and read this as "four does not belong to A" or any of the variants suggested above.

A set A is a **subset** of, or is contained in, a set B if for each $x \in A$, we also have $x \in B$. In this case we write $A \subseteq B$ or $B \supseteq A$. If both $A \subseteq B$ and $B \subseteq A$, the sets A and B have exactly the same elements and we write $A = B$. We reserve the symbol, $\emptyset$, for the empty set, i.e., the set with no elements whatever, and observe the following universal truths: $A \subseteq A$, $A = A$, $\emptyset \subseteq A$, and $A \subseteq \emptyset$ if and only if $A = \emptyset$. We also note that "$\subseteq$" is a transitive relation, i.e., $A \subseteq B$ and $B \subseteq C$ imply $A \subseteq C$.

The notation $A \subset B$, which means $A \subseteq B$ and $A \neq B$, is also encountered occasionally, though we shall have little if any use for it. To deny that one set is contained in another, for example, to assert that $A \subseteq B$ is false, we write $A \nsubseteq B$.

The student is warned here for the first time, as he will be a number of times as he proceeds through the book, that mathematical notation is highly variable. This is the case with the notation for set inclusion, namely "$\subseteq$," and proper set inclusion, namely "$\subset$." There are authors who use simply the symbol "$\subset$" for set inclusion and never introduce a notion of proper set inclusion. In reading other works the student should thus check on the ways in which the various notations are defined.

EXERCISES

0.1. Prove the universal truths mentioned above, i.e., for any set A each of the following is true:

(a) $A \subseteq A$.
(b) $A = A$.

(c) $\emptyset \subseteq A$.
(d) $A \subseteq \emptyset$ if and only if $A = \emptyset$.

0.2. For any sets A, B, and C, prove the following:
  (a) If $A \subseteq B$ and $B \subseteq C$, then $A \subseteq C$.
  (b) If $A \subset B$ and $B \subset C$, then $A \subset C$.

§3 The Algebra of Sets

We now define two operations on sets. The first is the **union** of two sets, written $A \cup B$, and defined by $A \cup B = \{x \mid x \in A \text{ or } x \in B\}$. The "or" is used in its nonexclusive sense, so that an element which is in both A and B is still in $A \cup B$. The second operation on two sets is the **intersection** operation, written $A \cap B$, and defined by $A \cap B = \{x \mid x \in A \text{ and } x \in B\}$. The following rules of operations on sets now follow quite straightforwardly from the definitions:

1. (a) $A \cup (B \cup C) = (A \cup B) \cup C$.
   (b) $A \cap (B \cap C) = (A \cap B) \cap C$.
2. (a) $A \cup B = B \cup A$.
   (b) $A \cap B = B \cap A$.
3. (a) $A \cap (B \cup C) = (A \cap B) \cup (A \cap C)$.
   (b) $A \cup (B \cap C) = (A \cup B) \cap (A \cup C)$.
4. (a) $A \cup A = A$.
   (b) $A \cap A = A$.
5. (a) $A \subseteq A \cup B$.
   (b) $A \supseteq A \cap B$.
6. (a) $A \subseteq C$ and $B \subseteq C$ imply $A \cup B \subseteq C$.
   (b) $A \supseteq C$ and $B \supseteq C$ imply $A \cap B \supseteq C$.
7. (a) $A \cup \emptyset = A$.
   (b) $A \cap \emptyset = \emptyset$.

As remarked earlier $A = B$ if and only if $A \subseteq B$ and $B \subseteq A$. Thus, in order to prove most of the above statements we must prove two inclusions. To demonstrate this we prove 2(a). Let $x \in A \cup B$, then $x \in A$ or $x \in B$. Consequently $x \in B$ or $x \in A$, whence $x \in B \cup A$; and it follows that $A \cup B \subseteq B \cup A$. Let $x \in B \cup A$, then $x \in B$ or $x \in A$. Consequently $x \in A$ or $x \in B$, whence $x \in A \cup B$; and it follows that $B \cup A \subseteq A \cup B$. From the two inclusions just proved it follows that $A \cup B = B \cup A$.

We shall in general assume that the sets we are discussing are all subsets of some universal set, U, and then we define the **complement** of A, which we designate by A^c, by $A^c = \{x \mid x \in U \text{ and } x \notin A\}$. The **difference** of two sets A and B we define as $A - B = \{x \mid x \in A \text{ and } x \notin B\}$. The further list of rules of operation which follow from these definitions is:

8. (a) $\varnothing^c = U$.
   (b) $U^c = \varnothing$.
9. (a) $A \cup A^c = U$.
   (b) $A \cap A^c = \varnothing$.
10. $(A^c)^c = A$.
11. If $A \subseteq B$, then $A^c \supseteq B^c$ and conversely.
12. (a) $(A \cup B)^c = A^c \cap B^c$.
    (b) $(A \cap B)^c = A^c \cup B^c$.
13. $A - B = A \cap B^c$.

Rules 12(a) and (b) above are called **DeMorgan's rules**; we shall encounter more general forms of these two shortly. The student should convince himself of the truth of the above rules by proving each of them. Once again the student is warned that the notation used for complements of sets, namely A^c, differs from what he is likely to find in other texts.

The list we now have is probably sufficient for our needs; however, a set of exercises follows which gives a number of further relations which may occasionally be useful.

EXERCISES

Prove each of the following, where A, B, and C designate sets in some universal set U.

0.3. $A \subseteq B$ does not in general imply $B \subseteq A$.

0.4. $A \subseteq B$ if and only if $A \cup B = B$.

0.5. $A \subseteq B$ if and only if $A \cap B = A$.

0.6. $A \cap (B - C) = B \cap (A - C) = (A \cap B) - C = (A \cap B) - (A \cap C)$.

0.7. $A - B = A - (A \cap B)$.

0.8. $A \cap B = \emptyset$ if and only if $A \subseteq B^c$.

0.9. If $A \cup B = U$ and $A \cap B = \emptyset$, then $B = A^c$.

0.10. $(A - B)^c = B \cup A^c$.

0.11. $(A - B) \cup (A - C) = A - (B \cap C)$.

0.12. $(A - C) \cup (B - C) = (A \cup B) - C$.

0.13. $(A - B) \cup (B - A) = (A \cup B) - (A \cap B)$.

0.14. $A - (A - B) = A \cap B$.

0.15. $A \cup (B - A) = A \cup B$.

0.16. $A \cap (B - A) = \emptyset$.

It is clear that by using the associative law for unions (or intersections), 1(a) and 1(b) above, we may take the union (or intersection) of any finite collection of sets. It is useful, however, to be able to take unions and intersections of arbitrary families of sets, not merely finite families. In order to define such a notion we have recourse to an indexing set, or as it is also called, a set of indices. We can think of an indexing set as a set of tags, so that each member of a family of sets has associated with it a tag from the indexing set. Thus, let A be a set of indices, and with each $\alpha \in A$ associate a set B_α, then define

$$\bigcup_{\alpha \in A} B_\alpha = \{x \mid x \in B_\alpha \text{ for some } \alpha \in A\}$$

and

$$\bigcap_{\alpha \in A} B_\alpha = \{x \mid x \in B_\alpha \text{ for all } \alpha \in A\}.$$

If A is the empty set, we define

$$\bigcup_{\alpha \in A} B_\alpha = \emptyset \quad \text{and} \quad \bigcap_{\alpha \in A} B_\alpha = U.$$

The following rules of operation then follow:

14. If $\alpha \in A$, then $\bigcap_{\alpha \in A} B_\alpha \subseteq B_\alpha \subseteq \bigcup_{\alpha \in A} B_\alpha$, $A \neq \emptyset$.
15. (a) If for each $\alpha \in A$, $B_\alpha \subseteq C$, then $\bigcup_{\alpha \in A} B_\alpha \subseteq C$.
    (b) If for each $\alpha \in A$, $B_\alpha \supseteq C$, then $\bigcap_{\alpha \in A} B_\alpha \supseteq C$.

16. If for each $\alpha \in A$ ($A \neq \emptyset$), $B_\alpha \subseteq C_\alpha$, then $\bigcup_{\alpha \in A} B_\alpha \subseteq \bigcup_{\alpha \in A} C_\alpha$ and $\bigcap_{\alpha \in A} B_\alpha \subseteq \bigcap_{\alpha \in A} C_\alpha$.

17. (a) $\bigcup_{\alpha \in A} (B_\alpha \cup C_\alpha) = (\bigcup_{\alpha \in A} B_\alpha) \cup (\bigcup_{\alpha \in A} C_\alpha)$.

   (b) $\bigcap_{\alpha \in A} (B_\alpha \cap C_\alpha) = (\bigcap_{\alpha \in A} B_\alpha) \cap (\bigcap_{\alpha \in A} C_\alpha)$.

18. $\bigcap_{\alpha \in A} (C \cup B_\alpha) = C \cup (\bigcap_{\alpha \in A} B_\alpha)$ for $A \neq \emptyset$.

19. $\bigcup_{\alpha \in A} (C \cap B_\alpha) = C \cap (\bigcup_{\alpha \in A} B_\alpha)$ for $A \neq \emptyset$.

20. (a) $(\bigcup_{\alpha \in A} B_\alpha)^c = \bigcap_{\alpha \in A} B_\alpha^c$.

    (b) $(\bigcap_{\alpha \in A} B_\alpha)^c = \bigcup_{\alpha \in A} B_\alpha^c$.

   $\Bigg\}$ (DeMorgan's rules)

As an alternative notation for unions and intersections of the type just discussed, we occasionally do the following: Let $\mathfrak{F}$ be a family of sets $\{B\}$, then $\bigcup_{B \in \mathfrak{F}} B = \{x \mid x \in B \text{ for some } B \in \mathfrak{F}\}$ and similarly for intersections.

We find occasional use for the following notation: if $\mathcal{A}$ is a family of sets $\{A\}$ and X is some fixed set then $\mathcal{A} \cap X$ means the family $\{A \cap X \mid A \in \mathcal{A}\}$.

Finally, let X_i, $i = 1, 2, \ldots, n$ be sets, then the **Cartesian product** of these sets $\underset{i=1}{\overset{n}{\times}} X_i = \{(x_1, x_2, \ldots, x_n) \mid x_i \in X_i\}$. This amounts to saying that the Cartesian product of the n sets is just the set of all possible n-tuples where the ith member of each n-tuple is chosen from the set X_i. In case there are only two sets we frequently prefer to write $X_1 \times X_2$ in place of $\underset{i=1}{\overset{2}{\times}} X_i$. For example, if $A = \{1, 2\}$ and $B = \{a, b, c\}$, then

$$A \times B = \{(1, a), (1, b), (1, c), (2, a), (2, b), (2, c)\}.$$

We shall discuss Cartesian products at greater length at a later time.

EXERCISES

0.17. Verify that DeMorgan's rules [20(a) and (b) above] hold in case $A = \emptyset$, as well as in the case $A \neq \emptyset$.

0.18. The Cartesian product of the real line, R, with itself is the real plane. If the first quadrant is the set

$$\{(x, y) \mid x > 0,\, y > 0,\, x,\, y \text{ real}\},$$

write the first quadrant as a Cartesian product.

0.19. If $R^+ = \{x \mid x \text{ real},\, x > 0\}$ and $\mathcal{Q}$ is the family of all subsets of R^+, then if $B = [-1, 1] = \{x \mid x \text{ real } -1 \leq x \leq 1\}$ describe $\mathcal{Q} \cap B$.

§ 4 Euler-Venn Diagrams

In dealing with problems in set theory a picturization of the relations between the sets is frequently helpful. Such a picture or diagram is usually called a Venn diagram, or Euler diagram, or even **Euler-Venn diagram**. Usually the universal set is pictured as a large rectangle and the pertinent sets are designated by means of regions inside the rectangle. For example, an Euler-Venn diagram which pictures $A \cap B$ is shown in Fig. 0.1.

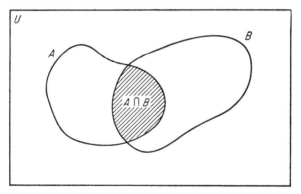

Figure 0.1

It should be noted that the diagram itself does not prove a particular result in set theory; however, it frequently gives some hint as to how a proof can be constructed. This is demonstrated, for example, in Figs. 0.2 and 0.3 below which suggest an approach to the proof of the equality $A \cap (B \cup C) = (A \cap B) \cup (A \cap C)$.

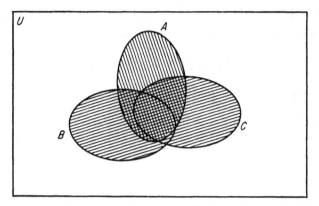

Figure 0.2 [Indicating $A \cap (B \cup C)$]

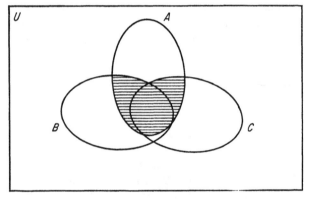

Figure 0.3 [Indicating $(A \cap B) \cup (A \cap C)$]

EXERCISES

0.20. Draw an Euler-Venn diagram to illustrate each of the following situations:
    (a) $A \subseteq B$.
    (b) $A \cap B \neq \emptyset$.
    (c) $A \subset B^c$.
    (d) $A \cap B \neq \emptyset, B \cap C \neq \emptyset, C \cap D \neq \emptyset, D \cap A \neq \emptyset,$
        $A \cap C = \emptyset, B \cap D = \emptyset$.

§ 5 Relations

We shall not study relations in general but confine our attention to two types of relations of particular utility for the sequel. First, however, we define the notion of relation in general. A **relation** in a set A is simply a collection of ordered pairs, (a, b), where $a \in A$, $b \in A$. We usually designate a relation by R, and write aRb to indicate that the ordered pair (a, b) is an element of the collection which determines the relation.

An **equivalence relation** is a relation which has the following three properties:

(1) aRa for each $a \in A$ **(reflexivity)**.
(2) If aRb then also bRa **(symmetry)**.
(3) If aRb and bRc, then also aRc **(transitivity)**.

If R is an equivalence relation in a set A, then we designate by $R(a)$ or $[a, R]$ or $[a]$ the set, $\{b \mid b \in A, aRb\}$, and we call this set, $[a]$, the **equivalence class** determined by a. If by a **partition** of a set, A, we mean a family, $\mathfrak{F}$, of subsets of A which are disjoint (i.e., $B, C \in \mathfrak{F}, B \neq C$ implies $B \cap C = \varnothing$) and whose union is A, then the following important theorem results:

0.1. Theorem. *Each equivalence relation in a set A induces a partition of A into equivalence classes. Conversely each partition, $\mathfrak{F}$, of A induces an equivalence relation in A whose equivalence classes are precisely the sets of $\mathfrak{F}$.*

Proof. Let $\mathfrak{F} = \{[a] \mid a \in A\}$. For each $a \in A$, $a \in [a]$, thus

$$\bigcup_{[a] \in \mathfrak{F}} [a] = A.$$

Now suppose $[a] \cap [b] \neq \varnothing$, then we can select $c \in [a] \cap [b]$, whence by definition of $[a]$ and $[b]$ we have cRa and cRb and by the symmetry and transitivity of R we have aRb. Let $d \in [a]$, then dRa, and since we already have aRb, it follows from the transitivity of R that dRb, whence $d \in [b]$. Similarly for any $e \in [b]$, we have also $e \in [a]$, whence $[a] = [b]$. Consequently the equivalence classes are either disjoint or they coincide, and $\mathfrak{F}$ is a partition of A.

Conversely let $\mathfrak{F}$ be a partition of A. Define aRb if and only if a and b belong to the same set $B \in \mathfrak{F}$. It then follows simply that R is an equivalence relation and the equivalence classes of R are precisely the sets of $\mathfrak{F}$. The details are left to the reader. ∎

Order relations are the second sort of relations we shall need to consider. In general, **order relations** are relations which are transitive; however, we distinguish two particular types of order relations which we shall need. The first of these is a **partial order relation**. We define a partial order relation in a set S to be a relation which satisfies the following conditions:

(1) For any $x \in S$, xRx.
(2) For any $x, y \in S$, xRy and yRx together imply $x = y$.
(3) For any $x, y, z \in S$, xRy and yRz together imply xRz.

A typical example of a partial order relation is the relation $A \subseteq B$ in the family of all subsets of some universe U.

The second order relation we shall need is the **simple order relation**. For our purposes we define a simple order relation as a partial order relation in which the further condition holds:

(4) If $x, y \in S$, and $x \neq y$ then either xRy or yRx.

A typical example of a simple order relation is the relation $x \leq y$ in the set of all real numbers.

The student is warned that he may encounter definitions of simple order relations different from the one above. In particular the following occurs occasionally as a definition of a simple order relation:

A simple order relation in a set S is a relation in which the following conditions hold:

(1) For any $x, y \in S$, one and only one of the following three hold: xRy, yRx, $x = y$.
(2) For any $x, y, z \in S$, xRy and yRz together imply xRz.

A typical example of this definition of simple order relation is the relation $x < y$ in the set of all real numbers. For our purposes, however, we shall prefer the previously given definition of simple order relation.

EXERCISES

0.21. In the set Z^+ of positive integers, let xRy mean $x - y$ is divisible by 7 (i.e., $x - y = 7k$, where k is an integer). Show that R is an equivalence relation and describe the equivalence classes.

0.22. In the set Z^+ of positive integers, let $m \leqq n$ mean m divides n (i.e., $n = mk$, where $k \in Z^+$). Show that "$\leqq$" is a partial ordering for Z^+.

§ 6 Infinite Sets

We shall define and discuss functions at somewhat greater length at the beginning of Chapter 2, but for the moment we require a particular type of function. We define a **one-to-one correspondence** (or function) between A and B in the following way: first think of f (the function or correspondence) as a set of ordered pairs (a, b), with $a \in A$, $b \in B$, such that if (a, b) and $(a, b') \in f$, then $b = b'$. This says simply that f is single valued, for we think of f as assigning the value $b \in B$ to the point (or element) $a \in A$. We frequently write $f(a) = b$, which notation should be familiar to the student from his study of calculus. Furthermore we insist that each $a \in A$ appear as a first element in some ordered pair $(a, b) \in f$. This then assures us that f assigns to each point (or element) $a \in A$ one and only one value $b \in B$. Under these circumstances f is called a function. In order that f be a one-to-one correspondence we insist further that each $b \in B$ appear as a second element in some ordered pair $(a, b) \in f$, and further that if (a, b) and $(a', b) \in f$, then $a = a'$. What this amounts to is that only one element of a has as its value, $f(a)$, the element $b \in B$, and that each element $b \in B$ does occur as a value of some $a \in A$.

The net effect of a one-to-one correspondence is that it assigns to each $a \in A$ exactly one $b \in B$ and to each $b \in B$ exactly one $a \in A$. A typical example of a one-to-one correspondence is the function $f(x) = x + 1$, where A and B are both the set of real numbers. An example of a function which is not a one-to-one correspondence is the function $f(x) = x^2$ in which A is the set of all real numbers, while B is the set of nonnegative real numbers.

A set is said to be **infinite** if there exists a one-to-one correspondence between the set Z^+ of positive integers and some subset of the given set. The set is said to be **countably infinite** if there is a one-to-one correspondence between Z^+ and the set itself. A set which is not infinite is said to be **finite**, and a set which is either finite or countably infinite is said to be **countable**. The word "denumerable" is frequently seen as a synonym for countable.

The following are easy consequences of these definitions: if A is infinite and $A \subseteq B$, then B is infinite; if B is finite and $A \subseteq B$, then A is finite; if A is infinite, and $a \in A$, then $A - \{a\}$ is infinite.

We now introduce an axiom which will make possible many of the proofs that follow. They may be possible without the assumption of this axiom, but at present this seems most unlikely. It is known that the assumption of this additional axiom, the so-called axiom of choice, does not introduce into the system we already have any inconsistencies that are not already there. The axiom is this:

Axiom of Choice. *Let $\mathfrak{F} = \{B_\alpha \mid \alpha \in A, A$ an indexing set$\}$ be a nonempty family of nonempty disjoint sets; then there exists a set C, such that $C \cap B_\alpha$ is a single element for each $\alpha \in A$.*

We can think of the single element in $C \cap B_\alpha$ as b_α, and we can define a function from A to C by defining $f(\alpha) = b_\alpha$. Under these circumstances f is called a choice function, for what it does for us is to choose an element (and only one) from each B_α. It is, in fact, the case that there are many equivalent statements of the axiom of choice, but we are not interested in pursuing these here. Later we shall need one other form of the axiom, one that is usually called Zorn's lemma, but we shall state this at the appropriate place. The interested student is referred to the references at the end of this chapter for suitable texts in which he can pursue this topic further.

The following are now simple consequences of the definition of finite and infinite sets:

Let F be a finite set, and let Z_n be the set of integers $\{k \mid 0 < k \leq n\}$, then either $F = \emptyset$ or there exists a one-to-one correspondence between Z_n and F for some integer n. Conversely, if either $F = \emptyset$ or there exists a one-to-one correspondence between F and Z_n for some integer n, then F is finite.

From this last remark it follows easily that the union of two finite sets is again finite, and that if A is infinite and F finite then $A - F$ is again infinite.

We come now to some results which we take a bit more care to prove in detail. Before we begin we assume that the reader is familiar with the real number system, and in particular is aware that the set of positive integers is **well ordered,** i.e., has the property that every nonempty set of positive integers has a smallest member.

Our first result says roughly that in the family of sets the smallest sets are the countable sets. If a finer classification is desired, we can say that the finite sets are the smallest, and are followed by the countably infinite sets. We prove

0.2. Theorem. *Any subset of a countable set is again countable.*

Proof. Let A be a countable set, and let $B \subseteq A$. If B is finite we are finished, so we assume that B is infinite. Let f be the one-to-one correspondence between A and Z^+, the set of positive integers, where we indicate the element of A which corresponds to the integer n by $f(n)$ in the usual functional notation. Let n_1 be the least integer such that $f(n_1) \in B$—such an integer exists because of the well ordering of the positive integers. Then let n_2 be the smallest integer such that $f(n_2) \in B - \{f(n_1)\}$, and inductively let n_k be the smallest integer such that $f(n_k) \in B - \bigcup_{i=1}^{k-1} \{f(n_i)\}$. We observe that the latter set, $B - \bigcup_{i=1}^{k-1} \{f(n_i)\}$, is nonempty for each k, otherwise we would have constructed a one-to-one correspondence between the set of positive integers Z_k and B which would imply that B is finite. We then define the one-to-one correspondence between Z^+ and B by $g(k) = f(n_k)$. The existence of this correspondence shows that B is countable. ∎

It will be important for us to know that the union of a countable family of countable sets is again countable. This result is stated in the following:

0.3. Theorem. *Let A be a countable indexing set, and with each $\alpha \in A$ let there be associated a countable set B_α, then the set $\bigcup_{\alpha \in A} B_\alpha$ is again countable.*

Remarks Preliminary to the Proof. Since we know that subsets of countable sets are countable, we might as well assume that each of the sets B_α as well as the set A is countably infinite, for if they are not (i.e., if they are finite in some cases) we can fill out the finite sets with further elements so that they are countably infinite. Then the union in which we are interested will be a subset of the union we shall prove countable and will itself be countable as a subset of a countable set. Similarly we may as well assume that the sets B_α are disjoint (i.e.,

$B_\alpha \cap B_\beta = \emptyset$ for $\alpha \neq \beta$), since if they are not we may think of the elements of all the B_α as distinct, form their union, keeping any like elements distinct, prove this set countable, and then think of the union in which we are interested as a subset of the union in which like elements have been kept distinct. Again, since countable sets have at worst countable subsets, we shall be finished.

Proof. By the remarks just made we assume A and each B_α as countably infinite, and the B_α disjoint. Since A is countably infinite there exists a one-to-one correspondence f between Z^+ and A, thus to each $\alpha \in A$ there corresponds a unique integer such that $f(n) = \alpha$. Let us rename the sets B_α by replacing the index α by the index n, where, of course, $f(n) = \alpha$ to obtain the family of sets B_n, $n = 1, 2, \ldots$. Now since each B_n is countably infinite there exists for each n a one-to-one correspondence f_n between Z^+ and B_n, so that for each $b \in B_n$ there exists a unique integer k such that $f_n(k) = b$. We tag the element b with the two indices n and k, writing it as b_{nk}. In this fashion each element in the union $\bigcup_{\alpha \in A} B_\alpha$ has attached to it a unique pair of indices. The following diagram makes this clear:

$$B_1: \quad b_{11} \quad b_{12} \quad b_{13} \quad \ldots$$
$$B_2: \quad b_{21} \quad b_{22} \quad b_{23} \quad \ldots$$
$$B_3: \quad b_{31} \quad b_{32} \quad b_{33} \quad \ldots$$

We now set up the following one-to-one correspondence:

1	2	3	4	5	6	7	8	9	10	11	12	13	14	15	16
↕	↕	↕	↕	↕	↕	↕	↕	↕	↕	↕	↕	↕	↕	↕	↕
b_{11}	b_{12}	b_{21}	b_{13}	b_{22}	b_{31}	b_{14}	b_{23}	b_{32}	b_{41}	b_{15}	b_{24}	b_{33}	b_{42}	b_{51}	b_{16}

and so on, in which we count first the elements whose indices sum to 2, then those whose indices sum to 3, then those whose indices sum to 4, etc., and for each such sum (i.e., 2, 3, 4, ...) we arrange the elements so that the first index occurs in the natural order for the integers. We thus establish the existence of a one-to-one correspondence between Z^+ and

$\bigcup_{\alpha \in A} B_\alpha$, and this proves that the union of countably many countable sets is again countable. ∎

It might appear that there is nothing worse than a countably infinite set, i.e., that every infinite set is necessarily countably infinite. Such, however, is not the case; and we give a classic proof, due to Cantor, to show that the real numbers between zero and one are not countable. We suppose that in fact the set of real numbers in the interval (0, 1) is countable and assume a one-to-one correspondence has been set up between Z^+ and these numbers. We indicate the correspondence by the following diagram:

$$1 \leftrightarrow .a_{11}a_{12}a_{13}a_{14}\ldots$$
$$2 \leftrightarrow .a_{21}a_{22}a_{23}a_{24}\ldots$$
$$3 \leftrightarrow .a_{31}a_{32}a_{33}a_{34}\ldots$$
$$4 \leftrightarrow .a_{41}a_{42}a_{43}a_{44}\ldots$$
$$\vdots \qquad \vdots \qquad \vdots$$

where each a_{ij} represents a digit, i.e., $0 \leq a_{ij} \leq 9$, and where it is assumed that where we have two alternate choices for the decimal expression of the real number, as for example in the case where $\frac{2}{10}$ could be written either as .2000... or as .1999..., we always choose the one that ends in a string of zeros. Now this one-to-one correspondence is such that to every positive integer there corresponds some real number in (0, 1) and conversely to each real number in (0, 1) there corresponds some integer. Consequently the infinite list of decimals given above is complete in the sense that every real number of (0, 1) occurs somewhere in the list. If, then, we can produce a real number between zero and one which is not in this list we shall have a contradiction, and this is precisely what we set out to do. We define $b = .b_1b_2b_3\ldots$ as follows: if a_{ii} is 5 let $b_i = 6$, if $a_{ii} \neq 5$ let $b_i = 5$. Now it is clear that b is not equal to any one of the decimals in our list for it differs from the nth one at the nth place. Also it is clear that $\frac{5}{9} \leq b \leq \frac{2}{3}$, so that $b \in (0, 1)$. This contradiction then shows that there cannot exist such a one-to-one correspondence between Z and (0, 1), and since (0, 1) is clearly infinite, containing as it does the set $\left\{\frac{1}{n} \mid n = 2, 3, \ldots\right\}$, the set of real numbers in (0, 1) is uncountable.

EXERCISES

0.23. Prove that if A is infinite, and $a \in A$, then $A - \{a\}$ is infinite.

0.24. Criticize the following "proof" that the set of positive integers, Z^+, is uncountable: Write each positive integer in the usual way, but precede the first digit by an infinite string of zeros stretching off to the left, e.g., 17 is written ...00017. Assume the set Z^+ is countable and set up the obvious one-to-one correspondence $n \leftrightarrow \ldots 000n$, e.g., 124 corresponds with ...000124. Write the list of these down in the usual order, thus:

$$\ldots 0001$$
$$\ldots 0002$$
$$\ldots 0003$$
$$\text{etc.}$$

Now construct a new number as follows: Reading down the diagonal entries in the above list, if the digit at the nth place in the nth number of the list is 5, let the entry in the new number at the nth place be 6, if the nth digit in the nth number of the list is different from 5, let the entry at the nth place of the new number be 5. The number thus constructed is different from the nth number in the list at the nth place, thus is different from all of them, and the one-to-one correspondence above constructed is not as claimed a one-to-one correspondence, whence the set Z^+ is uncountable.

§7 Miscellaneous Assumptions Regarding the Real Numbers

We assume that from his previous work the student is familiar with the real number system. In particular we assume that he is familiar with the principle of **mathematical induction** in either of the following two forms:

1. If S is a set of positive integers such that
    (a) $1 \in S$,
    (b) $k \in S$ implies $k + 1 \in S$,
    then $S = Z^+$, the set of all positive integers.
2. If S is a set of positive integers such that

(a) $1 \in S$,
(b) for each $k < n$, $k \in S$ implies $n \in S$,

then $S = Z^+$, the set of all positive integers.

We also assume that the set of positive integers is well ordered, i.e., each nonempty subset of Z^+ contains a smallest number—smallest in the sense of the usual ordering of the real numbers.

If T is a set of real numbers we say that t is an **upper bound** for T if $s \leq t$ for each $s \in T$, and u is a **lower bound** for T if $u \leq s$ for each $s \in T$. We say that a is a **least upper bound (supremum)** for T if a is an upper bound, and if b is any upper bound for T, then $a \leq b$; and we write sup $T = a$. Similarly c is a **greatest lower bound (infimum)** for T if c is a lower bound, and if d is any lower bound, then $d \leq c$; and we write inf $T = c$. We assume that any set of real numbers that has an upper bound has a supremum, and any set of real numbers that has a lower bound has an infimum.

We use the usual designation for intervals of real numbers, as follows:

$$(a, b) = \{x \mid x \text{ real}, a < x < b\}$$
$$[a, b) = \{x \mid x \text{ real}, a \leq x < b\}$$
$$(a, b] = \{x \mid x \text{ real}, a < x \leq b\}$$
$$[a, b] = \{x \mid x \text{ real}, a \leq x \leq b\}$$

where a and b are real numbers. Also we feel free to use the notation (a, ∞), (∞, b), $[a, \infty)$, and $(\infty, b]$ with the obvious meaning, e.g.,

$$(a, \infty) = \{x \mid x \text{ real}, a < x\}.$$

REFERENCES

Although there is a general list of references at the end of the book, there is listed here a set of books dealing principally with set theory and related matters.

LOGIC

1. Christian, R. R., *Introduction to Logic and Sets* (Preliminary ed.; Boston: Ginn, 1958).

2. Rosser, J. B., *Logic for Mathematicians* (New York: McGraw, 1953).

3. Kleene, S. C., *Introduction to Metamathematics* (Princeton, N.J.: Van Nostrand, 1952).

SETS

1. Bourbaki, N., *Théorie des Ensembles* (Paris: Actualités Scientifiques et Industrielles, Herman et Cie., 846 = 1141, 1951).

2. Halmos, P. R., *Naive Set Theory* (Princeton, N.J.: Van Nostrand, 1960).

3. Kamke, E., *Theory of Sets* (New York: Dover, 1950).

4. Suppes, P., *Axiomatic Set Theory* (Princeton, N.J.: Van Nostrand, 1960).

REAL NUMBER SYSTEM

1. Landau, E., *Foundations of Analysis* (New York: Chelsea, 1951).

HISTORICAL

1. Cantor, G., *Contributions to the Founding of the Theory of Transfinite Numbers* (New York: Dover, n.d.).

CHAPTER 1

TOPOLOGICAL SPACES—BASIC DEFINITIONS AND THEOREMS

§ 1 Neighborhood Systems and Topologies

As we look back over the development of mathematics through the ages, we seem to discern a dichotomy. On the one hand there is algebra, which has studied the arithmetic structure of various mathematical systems, as for example in algebra proper, both classical and modern, number theory, and the like. On the other hand there is geometry, which has studied the geometric or spatial characteristics of mathematical systems, as for example in geometry proper, both Euclidean and non-Euclidean, differential geometry, and topology. This dichotomy is more apparent than real, for in fact, all mathematics is related, and the two fields mentioned above are inextricably intertwined, so that as one advances in the study of either field, notions from the other become essential. In fact, it is probably the case that the greatest synthesis in mathematics takes place in the field of analysis, in which the notions of both algebra and topology are simultaneously brought to bear on the study of mathematical systems.

Our purpose here is to study what is ordinarily called point set topology. Point set topology forms the background for a number of more advanced topics. Of these advanced topics we single out three, and we have in mind laying a foundation that will enable the student to go on in any of these three. The three we have in mind are *analysis*, which we have already mentioned, *advanced point set topology*, which will extend the ideas we study here, and *algebraic topology*, which utilizes some of the techniques of algebra to further the study of topological spaces.

Since our concern in point set topology will be with the spatial characteristics of a set, we shall be thinking largely in terms of geometric notions, and we shall thus call the elements of the set we are studying, points. We introduce axioms into this set of points via definitions. Now one of our principal concerns in geometry (or topology) is the notion of "closeness." We have, in some way, to say which points lie "close" to other points. We can introduce a notion of closeness in a variety of different ways, but we seek one that is quite general, so that by specialization we can obtain many of the systems we already know well. The method we select defines first the notion of a neighborhood system of a point x of the set we are studying. In a certain sense, a neighborhood of a point x is a set of points which lie "close" to the point. Let us get under way and define a neighborhood system, via the following:

1.1. Definition. *Let X be a set, and for each point $x \in X$, let $\mathfrak{U}_x = \{U(x)\}$ be a nonempty family of subsets of X associated with x, such that*

(1) $x \in U(x)$ *for each* $U(x) \in \mathfrak{U}_x$.
(2) *If* $V \supseteq U(x)$ *for some* $U(x)$, *then* $V \in \mathfrak{U}_x$.
(3) *If* U *and* $V \in \mathfrak{U}_x$, *then* $U \cap V \in \mathfrak{U}_x$.
(4) *If* $U \in \mathfrak{U}_x$, *then there exists* $V \in \mathfrak{U}_x$ *such that if* $y \in V$ *then* $U \in \mathfrak{U}_y$.

*Then $\mathfrak{U}_x$ is called a **system of neighborhoods at x**.*

It is useful to observe at this point that the set V defined in (4) of the above definition is such that $V \subseteq U$. This is easily seen, for let $y \in V$, then $U \in \mathfrak{U}_y$, hence by (1) of 1.1 $y \in U$, whence $V \subseteq U$.

We now have some idea how we want points "close" to a given point (i.e., neighborhoods) to behave. Of course, our notion of neighborhood is very general and includes all sorts of strange cases; in particular we shall find (Ex. 1.6 below) that the whole set X is a neighborhood of each of its points and in fact need be the only neighborhood of each point. These pathological cases need not disturb us, for we shall make our future definitions in such a way that a given property holds for *every* neighborhood of a point, and thus most of the pathology disappears. Now we stick all the neighborhood systems of a set together to form a topological space. We do this via the following:

1.2. Definition. *Let X be a set, and let $\mathfrak{J} = \{\mathfrak{U}_x \mid x \in X\}$ be an assignment of neighborhood systems for each point $x \in X$, then the pair $(X, \mathfrak{J})$ is* called a **topological space**. $\mathfrak{I}$ is called the **topology** for the space $(X, \mathfrak{I})$, and as mentioned before the elements of X are called **points**. Further, if $\mathfrak{I} = \{\mathfrak{U}_x\}$ and $\mathfrak{I}' = \{\mathfrak{U}'_x\}$ are two topologies for a set X, then $\mathfrak{I} = \mathfrak{I}'$ if and only if $\mathfrak{U}_x = \mathfrak{U}'_x$ for each $x \in X$.

Actually the last sentence in the above definition is unnecessary, for since a topology is simply a collection of families of subsets of a set X, two topologies will be the same if and only if they are equal as sets, and this will be so if and only if $\mathfrak{U}_x = \mathfrak{U}'_x$ for each $x \in X$. It is well, however, to make this quite explicit, for it is quite possible that two topologies may be alike in that they specify the same subsets of X as neighborhoods of points of X, yet they will not be equal as topologies, since the assignment of individual neighborhoods to points of the space may be different. The following example points up this distinction: let $X = \{x, y, z\}$ and let $\mathfrak{I}$ be determined by

$$\mathfrak{U}_x = \{\{x\}, \{x, y\}, \{x, z\}, \{x, y, z\}\},$$
$$\mathfrak{U}_y = \{\{x, y, z\}\}$$

and

$$\mathfrak{U}_z = \{\{x, y, z\}\},$$

while $\mathfrak{I}'$ is determined by

$$\mathfrak{U}_x = \{\{x\}, \{x, y\}, \{x, z\}, \{x, y, z\}\},$$
$$\mathfrak{U}_y = \{\{x, y\}, \{x, y, z\}\}$$

and

$$\mathfrak{U}_z = \{\{x, y, z\}\}.$$

Now $\mathfrak{I}$ and $\mathfrak{I}'$ are alike in that they specify exactly the same sets of X as neighborhoods, but they are not the same topology, since in $\mathfrak{I}$, $\{x, y\}$ is a neighborhood of x but not of y, while in $\mathfrak{I}'$ $\{x, y\}$ is a neighborhood of both x and y.

It is frequently convenient to say simply that X is a topological space, rather than having to specify the topology, $\mathfrak{I}$, and having to write $(X, \mathfrak{I})$, since, more often than not, we are interested not so much in a particular topology, but rather in properties that any topology will possess. We shall thus feel free in what follows to omit any specific mention of the topology unless it is important to the context to emphasize a particular topology, or to distinguish between different topologies.

EXAMPLES (AND EXERCISES)

Every Example is an Exercise, and the statements made in the Examples should be verified as Exercises.

1.1. Let R be the real line, define $(a, b) = \{x \mid a < x < b\}$ and define

$$\mathfrak{U}_x = \{U \mid x \in (a, b) \subseteq U \text{ for some } a, b \in R, a < b\},$$

then $\mathfrak{U}_x$ is a neighborhood system at x and generates a topology for R according to 1.2. The topology so generated is called the **usual topology for R**.

NOTE: The function ρ of the following Ex. 1.2 and the function

$$\rho(f, g) = \int_0^1 |f - g| \, dx$$

implicitly defined in Ex. 1.4 are essentially metrics, which we shall discuss at length in Chapter 5. For the time being, however, the student should note and feel free to use these properties of metrics:

(a) $\rho(x, x) = 0$ for every point x in the space.
(b) $\rho(x, y) = \rho(y, x)$ for every x, y in the space.
(c) $\rho(x, y) \leq \rho(x, z) + \rho(z, y)$ for every x, y, z in the space.

1.2. Let $E = R \times R$, where R is the real line, i.e., E is the real plane. Let $x, y \in E$, where $x = (a, b)$, $y = (c, d)$ and define

$$\rho(x, y) = [(a - c)^2 + (b - d)^2]^{1/2}.$$

$\rho(x, y)$ is called the distance from x to y. Define

$$S_\epsilon(x) = \{y \mid \rho(x, y) < \epsilon\},$$

called an open ϵ-sphere about x, and finally define

$$\mathfrak{U}_x = \{U \mid U \supseteq S_\epsilon(x) \text{ for some } \epsilon > 0\}.$$

Then $\mathfrak{U}_x$ is a neighborhood system at x, and generates a topology according to 1.2. The topology so generated is called the **usual topology for the plane**.

1.3. Let X be a partially ordered set under some ordering, $\leq$, i.e., (α) $x \leq y$ and $y \leq z$ imply $x \leq z$, (β) $x \leq y$ and $y \leq x$ imply $x = y$, and (γ) $x \leq x$ for all $x \in X$. Define $S_r(x) = \{y \mid x \leq y\}$, and define $\mathfrak{U}_x = \{U \mid U \supseteq S_r(x)\}$, then $\mathfrak{U}_x$ is a neighborhood system at x. The topology, $\mathfrak{I}$, defined (via 1.2) for X is called the **right order topology** for X. A left order topology may also be defined by starting instead with the sets $S_l(x) = \{y \mid y \leq x\}$.

1.4. Let X be the set of all real-valued integrable functions whose common domain is the interval $[0, 1]$. For $f \in X$ define

$$S_\epsilon(f) = \{g \mid g \in X, \int_0^1 |f - g|\, dx < \epsilon\},$$

and define

$$\mathfrak{U}_f = \{U \mid U \supseteq S_\epsilon(f) \text{ for some } \epsilon > 0\},$$

then $\mathfrak{U}_f$ is a neighborhood system at f, and such neighborhood systems generate a topology by 1.2.

1.5. Let X be a set, and let $\mathfrak{U}_x = \{U \mid x \in U\}$ for each $x \in X$, then $\mathfrak{U}_x$ is a neighborhood system at x, and the topology thus generated is called the **discrete topology** for X.

1.6. Let X be a set, and let $\mathfrak{U}_x = \{X\}$ for each $x \in X$, then $\mathfrak{U}_x$ is a neighborhood system for x, and the topology thus generated is called the **trivial topology** for X.

1.7. Let X be a set, $\mathfrak{I}$ and $\mathfrak{I}'$ two topologies for X, then $\mathfrak{I} = \mathfrak{I}'$ if and only if for each $x \in X$ and for each $U \in \mathfrak{U}_x \in \mathfrak{I}$ there is a $U' \in \mathfrak{U}'_x \in \mathfrak{I}'$ such that $U' \subseteq U$, and conversely for each $V' \in \mathfrak{U}'_x \in \mathfrak{I}'$ there is a $V \in \mathfrak{U}_x \in \mathfrak{I}$ such that $V \subseteq V'$.

§ 2 Open Sets in a Topological Space

The next definition and two theorems introduce the class of so-called open sets of a topological space. These distinguished subsets of a space play an important role in the topology of the space.

1.3. Definition. *A set $O \subseteq X$, where X is a topological space, is said to be **open** provided $O \in \mathfrak{U}_x$ for each $x \in O$. The family of open sets of a space X is designated by $\mathfrak{O}$.*

1.4. Theorem. *Let X be a topological space, then U is a neighborhood of $x \in X$ (i.e., $U \in \mathfrak{U}_x$) if and only if there exists $O \in \mathfrak{O}$ (i.e., O is an open set of X) such that $x \in O \subseteq U$.*

Proof. (1) Let U be a neighborhood of x, and let

$$O = \{y \mid \text{there exists } W \in \mathfrak{U}_y, W \subseteq U\}.$$

We observe first that since $U \in \mathfrak{U}_x$ and $U \subseteq U$, we have $x \in O$. Now let $y \in O$, then there exists $W \in \mathfrak{U}_y$ such that $W \subseteq U$, by definition of O. Also, by 1.1(4), there exists a $V \in \mathfrak{U}_y$ such that $z \in V$ implies $W \in \mathfrak{U}_z$, and as already remarked $V \subseteq W$. We have thus that for each $z \in V$, $z \in W \subseteq U$ and $W \in \mathfrak{U}_z$ whence $z \in O$, and consequently $V \subseteq O$. We have, however, $V \in \mathfrak{U}_y$ and $V \subseteq O$, hence by 1.1(2), $O \in \mathfrak{U}_y$, and thus by 1.3, O is open.

(2) Let $x \in X, O \in \mathfrak{O}$, and $x \in O \subseteq U$. By 1.3, $O \in \mathfrak{U}_x$, and by 1.1(2), $U \in \mathfrak{U}_x$. ∎

1.5. Theorem. *Let X be a topological space, then*

(1) *The union of any number of open sets is open.*
(2) *The intersection of any two (consequently of any finite number) of open sets is open.*
(3) *X is open.*
(4) *$\emptyset$ is open.*

Proof. (1) Let A be an indexing set, and let O_α be open for each $\alpha \in A$. Further let $O = \bigcup_{\alpha \in A} O_\alpha$, and let $x \in O$, then $x \in O_\alpha$ for some α. Since O_α is open $O_\alpha \in \mathfrak{U}_x$, and since $O_\alpha \subseteq O$, $O \in \mathfrak{U}_x$ by 1.1(2), whence O is open by 1.3.

(2) Let O_1 and O_2 be open, $O = O_1 \cap O_2$, and let $x \in O$. Now $x \in O_1$ and $x \in O_2$, and since O_1 and O_2 are open $O_1 \in \mathfrak{U}_x$ and $O_2 \in \mathfrak{U}_x$, consequently by 1.1(3), $O = O_1 \cap O_2 \in \mathfrak{U}_x$, whence by 1.3, O is open.

(3) Let $x \in X$, then $\mathfrak{U}_x \neq \emptyset$; there exists $U \in \mathfrak{U}_x$. Since $U \subseteq X$, $X \in \mathfrak{U}_x$ by 1.1(2), and X is open by 1.3.

(4) If $\emptyset$ were not open, there would exist $x \in \emptyset$ such that $\emptyset \notin \mathfrak{U}_x$. Since, however, there is no such x ($\emptyset$ being empty), this is clearly false, hence $\emptyset$ is open. ∎

In introducing a topology into a set X, we started with the notion of neighborhood. This is only one of many avenues to the same goal;

in particular we might just as well have started with the notion of open set, and introduced the topology in a somewhat different way. That we reach the same goal is shown by the following:

1.6. Theorem. *Let X be a set, and let there be specified a family, $\mathfrak{O}$, of subsets of X, such that*

(1) *The union of any number of sets of $\mathfrak{O}$ is again in $\mathfrak{O}$.*
(2) *The intersection of any two (consequently of any finite number) of sets of $\mathfrak{O}$ is again in $\mathfrak{O}$.*
(3) $X \in \mathfrak{O}$.
(4) $\varnothing \in \mathfrak{O}$.

Then there is one and only one way in which to specify a topology, $\mathfrak{J}$, in X such that the sets of $\mathfrak{O}$ will be the open sets of the topology $\mathfrak{J}$. We say that the family $\mathfrak{O}$ generates $\mathfrak{J}$.

Proof. Define for each $x \in X$,

$$\mathfrak{U}_x = \{U \mid x \in O \subseteq U \text{ for some } O \in \mathfrak{O}\}.$$

We verify that $\mathfrak{U}_x$ is a neighborhood system for x.

(1) $\mathfrak{U}_x \neq \varnothing$ since by (3) $X \in \mathfrak{U}_x$. For each $U \in \mathfrak{U}_x$, $x \in O \subseteq U$, hence $x \in U$.
(2) Let $U \in \mathfrak{U}_x$, and $U \subseteq V$, then there exists $O \in \mathfrak{O}$, such that $x \in O \subseteq U \subseteq V$, and $V \in \mathfrak{U}_x$.
(3) If $U, V \in \mathfrak{U}_x$, then there exist $O_1, O_2 \in \mathfrak{O}$, such that $x \in O_1 \subseteq U$, $x \in O_2 \subseteq V$. By (2), $O_1 \cap O_2 \in \mathfrak{O}$, thus $x \in O_1 \cap O_2 \subseteq U \cap V$, and $U \cap V \in \mathfrak{U}_x$.
(4) Let $U \in \mathfrak{U}_x$, then there is an $O \in \mathfrak{O}$, such that $x \in O \subseteq U$. We remark that $O \in \mathfrak{U}_x$. Let $V = O$, and let $y \in V$, then $y \in V = O \subseteq U$; and $U \in \mathfrak{U}_y$.

Thus Definition 1.1 is satisfied, and $\mathfrak{U}_x$ is a neighborhood system for x.

We now verify that $\mathfrak{O}$ is the set of open sets generated by the topology $\mathfrak{J} = \{\mathfrak{U}_x \mid x \in X\}$. Let $O \in \mathfrak{O}$, and let $x \in O$, then $O \in \mathfrak{U}_x$ by definition of $\mathfrak{U}_x$, and consequently $O \in \mathfrak{U}_x$ for each $x \in O$, whence O is open in the topology $\mathfrak{J}$. Conversely, let U be open in the topology $\mathfrak{J}$, then $U \in \mathfrak{U}_x$ for each $x \in U$. By definition of $\mathfrak{U}_x$, there exists $O_x \in \mathfrak{O}$ such that

$x \in O_x \subseteq U$ for each $x \in U$. Clearly $U = \bigcup_{x \in U} O_x$, whence by hypothesis (1), $U \in \mathcal{O}$.

Finally we must show that the topology $\mathfrak{T}$ is uniquely determined. Since, however, by 1.4 U can be a neighborhood of x if and only if there exists an open set O, such that $x \in O \subseteq U$, it is clear that $\mathfrak{U}_x$ can be defined in no other way than the way chosen. ∎

Notice that $\mathcal{O}$ generates $\mathfrak{T}$ in a quite explicit fashion; namely, if we are given the family $\mathcal{O}$, we define for each $x \in X$

$$\mathfrak{U}_x = \{U \mid U \subseteq X, x \in O \subseteq U \text{ for some } O \in \mathcal{O}\}.$$

Any family $\mathcal{O}$ which satisfies the four conditions of 1.6 thus uniquely determines a topology $\mathfrak{T}$. We could if we wished have defined the topology of a space to be simply the family $\mathcal{O}$, and this is what is usually done. The student is alerted to this fact, so that when he reads other works on topology he will not be confused by alternate but equivalent definitions of a topology for a space. We have chosen the approach to a topology via neighborhood systems in order to retain as much of the geometric flavor as possible.

EXAMPLES (AND EXERCISES)

1.8. On the real line specify $\mathcal{O}$ to be the collection of all sets which are arbitrary unions of open intervals (a, b), $a < b$, then $\mathcal{O}$ generates the usual topology for the reals.

1.9. In the real plane specify $\mathcal{O}$ to be the collection of all sets which are arbitrary unions of open spheres, then $\mathcal{O}$ generates the usual topology for the plane.

1.10. In a partially ordered set X, specify $\mathcal{O}$ to be the family of all sets which are arbitrary unions of the sets $S_r(x)$, $x \in X$ (Cf. Ex. 1.3), then $\mathcal{O}$ generates the right order topology for X.

1.11. In the set of all integrable real-valued functions with common domain $[0, 1]$, specify $\mathcal{O}$ to be the collection of all sets which are arbitrary unions of sets of the form $S_\epsilon(f)$, $\epsilon > 0$, f an integrable real-valued function on $[0, 1]$, then $\mathcal{O}$ generates the topology of Ex. 1.4.

1.12. In any nonempty set X, specify $\mathcal{O}$ to be the family of all subsets of X, then $\mathcal{O}$ generates the discrete topology for X.

1.13. In any nonempty set X, specify $\mathcal{O}$ to be the family $\{X, \varnothing\}$, then $\mathcal{O}$ generates the trivial topology for X.

§ 3 Limit Points and the Derived Set

Since the study of analysis depends to some extent upon the structure of the space in which the functions being studied have their arguments and their values, and in this way depends upon some topological considerations, it is not surprising that topology should concern itself with some of the notions of analysis, in particular with some of the limiting processes. When we study limits of real functions, or limits of sequences, the limit of the function (or sequence) lies in a certain sense arbitrarily close to the values of the function (or sequence). We generalize this notion here by means of the notion of limit point, in the following:

1.7. Definition. *Let X be a topological space, and let $A \subseteq X$, the point $x \in X$ is said to be a **limit point** of A, provided that for each $U \in \mathfrak{U}_x$, $U \cap A$ contains a point $y \neq x$.*

EXAMPLES (AND EXERCISES)

1.14. On the real line with the usual topology, a as well as b is a limit point of the open interval (a, b).

1.15. In the real plane with the usual topology, any point of the form $(0, y)$ is a limit point of the set $D = \{(x, y) \mid x > 0\}$.

1.16. In the right order topology for a partially ordered set X, if $x \in X$ is not the least element of X (i.e., if it is false that $x \leq y$ for all $y \in X$), then any $y < x$ is a limit point of the set $S_r(x)$.

1.17. In the set of all integrable functions on $[0, 1]$ with the topology defined as in Ex. 1.4 let

$$Y = \{f \mid \int_0^1 f\, dx \neq 0\}$$

and let $g(x) = 0$ for $x \neq 1$, and $g(1) = 1$. Then g is a limit point of Y. In fact any function which has values different from zero for only finitely many values of $x \in [0, 1]$ is a limit point of Y.

1.18. Let X be a nonempty set with the discrete topology, let $A \subseteq X$ and let $x \in X$, then x is not a limit point of A.

1.19. Let X be a set with at least two elements, and let X have the trivial topology, let $A \subseteq X$ and let $x \in X$, then x is a limit point of A unless $A = \emptyset$ or $\{x\}$.

1.8. Definition. *Let X be a topological space and let $A \subseteq X$. The derived set of A, written A', is the set of all $x \in X$ such that x is a limit point of A.*

EXAMPLES (AND EXERCISES)

1.20. On the real line with the usual topology, let $A = (a, b)$, then $A' = [a, b]$. Let $B = \{x \mid 0 < x \leq 1 \text{ or } x = 2\}$, then $B' = [0, 1]$.

1.21. In the real plane with the usual topology, let $D = \{(x, y) \mid x > 0\}$, then $D' = \{(x, y) \mid x \geq 0\}$. Let $E = \{(x, y) \mid x \text{ and } y \text{ integers}\}$, then $E' = \emptyset$.

1.22. Let X be a partially ordered set with the right order topology, and let $A = \{x\}$, then $A' = \{y \mid y < x\}$.

1.23. In any topological space $\emptyset' = \emptyset$.

1.24. In any topological space, if $A \subseteq B$, then $A' \subseteq B'$.

§4 The Closure of a Set

The situation is now this: we begin with an arbitrary set A, in a topological space X, and derive from this set a new set, A', the set of all the limit points of A. It seems perhaps natural at this point to construct a further set, by throwing together into one set all the points both of A and of A', and this is precisely what we do in the following:

1.9. Definition. *Let X be a topological space, and let $A \subseteq X$. The closure of A, written $\overline{A}$, is the set $A \cup A'$.*

EXAMPLES (AND EXERCISES)

1.25. On the real line with the usual topology, let $A = (a, b)$, then $\overline{A} = [a, b]$. Let $B = \{x \mid 0 < x \leq 1 \text{ or } x = 2\}$, then

$$\overline{B} = \{x \mid 0 \leq x \leq 1 \text{ or } x = 2\}.$$

1.26. In the real plane with the usual topology, let $D = \{(x, y) \mid x > 0\}$, then $\overline{D} = \{(x, y) \mid x \geq 0\}$. Let $E = \{(x, y) \mid x \text{ and } y \text{ integers}\}$, then $\overline{E} = E$.

1.27. Let X be a simply ordered set with the right order topology, and let $A = S_r(x)$, then $\overline{A} = X$.

1.28. Let X be a nonempty set with the discrete topology, and let $A \subseteq X$, then $\overline{A} = A$.

1.29. Let X be a nonempty set with the trivial topology, and let $A \subseteq X$, then $\overline{A} = X$ unless $A = \varnothing$.

1.30. In any topological space $\overline{\varnothing} = \varnothing$.

1.31. In any topological space, X, $\overline{X} = X$.

1.32. In any topological space, X, let $A \subseteq X$, then $A \subseteq \overline{A}$.

1.33. In any topological space, X, $x \in \overline{A}$ if and only if for each $U \in \mathfrak{U}_x$, $U \cap A \neq \varnothing$.

The closure of a set A in a topological space X has, of course, a number of interesting properties. Among others is the property that it is maximal in the sense that no larger set is obtained from $\overline{A}$ by adding further limit points of $\overline{A}$ to it. This is shown by the following:

1.10. Theorem. *Let X be a topological space, and let $A \subseteq X$, then $\overline{\overline{A}} = \overline{A}$.*

Proof. By 1.9 $\overline{\overline{A}} = \overline{A} \cup \overline{A}'$. We show that $\overline{A}' \subseteq \overline{A}$. Let $x \in \overline{A}'$, and suppose $x \notin \overline{A} = A \cup A'$, then $x \notin A$ and further, since also $x \notin A'$, there exists some $U \in \mathfrak{U}_x$ such that $U \cap A = \varnothing$. Select O, open, such that $x \in O \subseteq U$, then $O \in \mathfrak{U}_x$, and further since $O \cap A \subseteq U \cap A = \varnothing$, we have $O \cap A = \varnothing$. Now since $x \in \overline{A}'$, $O \cap \overline{A}$ contains some point

$y \neq x$. Thus $y \in \overline{A}$, and since $O \cap A = \emptyset$, $y \in A'$. Since O is open, $O \in \mathfrak{U}_y$ by 1.3, thus there exists $z \neq y$ such that $z \in O \cap A$. This, however, contradicts $O \cap A = \emptyset$, consequently $x \in \overline{A}$. This completes the proof that $\overline{A}' \subseteq \overline{A}$. Finally $\overline{\overline{A}} = \overline{A} \cup \overline{A}' = \overline{A}$, since $\overline{A}' \subseteq \overline{A}$. ∎

There are also a number of relations between unions and intersections of sets and their closures, as well as certain special properties of a similar nature which depend not only upon the properties of intersections, unions, and closure, but also upon whether the sets under consideration are open. Several of the more important and useful results of this kind are found in the following two theorems.

1.11. Theorem. *Let X be a topological space and let $A \subseteq X$ and $B \subseteq X$, then*

(1) *If $A \subseteq B$, then $\overline{A} \subseteq \overline{B}$.*
(2) $\overline{A \cap B} \subseteq \overline{A} \cap \overline{B}$.
(3) $\overline{A \cup B} = \overline{A} \cup \overline{B}$.

Proof. (1) By Ex. 1.24 if $A \subseteq B$, then $A' \subseteq B'$, consequently $\overline{A} = A \cup A' \subseteq B \cup B' = \overline{B}$.

(2) Let $x \in \overline{A \cap B}$, and let $U \in \mathfrak{U}_x$, then $U \cap (A \cap B) \neq \emptyset$, consequently neither $U \cap A$ nor $U \cap B$ is empty, and $x \in \overline{A}$ and $x \in \overline{B}$, whence $x \in \overline{A} \cap \overline{B}$. Thus $\overline{A \cap B} \subseteq \overline{A} \cap \overline{B}$.

(3) Since $A \subseteq A \cup B$ and $B \subseteq A \cup B$, we have by part (1) of this theorem that $\overline{A} \subseteq \overline{A \cup B}$ and $\overline{B} \subseteq \overline{A \cup B}$, whence $\overline{A} \cup \overline{B} \subseteq \overline{A \cup B}$. Now let $x \in \overline{A \cup B}$, and suppose $x \notin \overline{A}$ and $x \notin \overline{B}$, then there exist U, $V \in \mathfrak{U}_x$ such that $U \cap A = \emptyset$ and $V \cap B = \emptyset$. Now $U \cap V \in \mathfrak{U}_x$ and

$$U \cap V \cap (A \cup B) = (U \cap V \cap A) \cup (U \cap V \cap B)$$
$$\subseteq (U \cap A) \cup (V \cap B) = \emptyset;$$

but this contradicts $x \in \overline{A \cup B}$. Consequently either $x \in \overline{A}$ or $x \in \overline{B}$, whence $x \in \overline{A} \cup \overline{B}$. Thus $\overline{A \cup B} \subseteq \overline{A} \cup \overline{B}$, and finally $\overline{A \cup B} = \overline{A} \cup \overline{B}$. ∎

1.12. Theorem. *Let X be a topological space, and let $A \subseteq X$ be open and let $B \subseteq X$, then $A \cap \overline{B} \subseteq \overline{A \cap B}$.*

Proof. Let
$$x \in \overline{A \cap B} = A \cap (B \cup B') = (A \cap B) \cup (A \cap B').$$
If $x \in A \cap B$, then
$$x \in (A \cap B) \cup (A \cap B)' = \overline{A \cap B}.$$
If $x \in A \cap B'$, then since $A \in \mathfrak{U}_x$, if $U \in \mathfrak{U}_x$, $A \cap U \in \mathfrak{U}_x$ by 1.1(3). Now since $x \in B'$, $(A \cap U) \cap B$ contains a point $y \neq x$. Thus for each $U \in \mathfrak{U}_x$, $U \cap (A \cap B) = (A \cap U) \cap B$ contains a point $y \neq x$, whence
$$x \in (A \cap B)' \subseteq \overline{A \cap B}.$$
In either case $x \in \overline{A \cap B}$, whence $A \cap \overline{B} \subseteq \overline{A \cap B}$. ∎

EXERCISE

1.34. Give an example on the real line in which A is an open set, and the sets $A \cap \overline{B}$, $\overline{A} \cap \overline{B}$, and $\overline{A \cap B}$ are all distinct. Give an example in which A is not open and $A \cap \overline{B} \not\subseteq \overline{A \cap B}$.

§ 5 Closed Sets

It is perhaps reasonable to ask at this point if there are sets which already contain all their limit points, and if such sets have any interesting and distinctive properties. Of course, we already know that there are sets which contain all their limit points, since by 1.10, $\overline{A}$ is such a set. We now set about investigating some of the properties of such sets, but in order to have a name for such sets we first introduce the following:

1.13. Definition. *Let X be a topological space, let $A \subseteq X$, then A is said to be **closed** provided $A = \overline{A}$.*

As innocuous as this definition looks, it is nonetheless the source of some difficulty. The choice of the word "closed" for the type of sets described in the definition is perhaps a poor one, for one intuitively has the notion that open and closed are antithetical notions, in that, for example, doors are either open or closed. Unfortunately this is not the case with sets in a topological space. It might be better to think of open and closed sets as being akin to the right to own a car. This is a right that is open to anyone, yet there are certain colleges that prohibit students from owning cars, so that to these students this right is closed, so that this right is at one and the same time open and closed. It is important to remember that this is the case with sets, namely that open and closed are not antithetical, that a set may be both open and closed at the same time, that it may be open but not closed, that it may be closed but not open, and that it may be neither open nor closed. Each of these possibilities is in fact realized on the real line with the usual topology, for the whole space, R, is both open and closed (as is also the empty set), the interval (a, b) $a < b$ is open but not closed, the interval $[a, b]$ is closed but not open, and the interval $[a, b)$ is neither open nor closed—these remarks should, of course, be verified by the reader.

We might say a few more words about closed sets before we go on to investigate their properties. First, pursuing the ideas in the preceding paragraph, if we are in a position to want to prove a set closed, it will do us no good whatever to prove it is not open, since we have already seen by example that these two properties can be possessed by the same set at the same time. It is, of course, equally true that it does no good to prove a set is not closed if our object is to prove that it is open.

It is also useful to observe that in any space the sets $\varnothing$ and X (the whole space) are invariably closed, and for that matter also invariably open. It is interesting to speculate at this point on what character a space must have if these two are the only sets which are at one and the same time open and closed—interesting, yes, but perhaps premature. [Cf. 4.3.]

It is, of course, quite pleasant that the closure of a set is in fact closed, as is shown by 1.10; and it is further rather nice that even though open and closed are not antithetical notions, nonetheless, open and closed sets are related as is shown in the following:

1.14. Theorem. *A set A, in a topological space X, is closed if and only if its complement, A^c, is open.*

Proof. Let A be closed, and let $x \in A^c$. Since $x \notin A = \overline{A}$, there exists a neighborhood $U \in \mathfrak{U}_x$ such that $U \cap A = \varnothing$, as is shown in Ex. 1.33. Consequently, $U \subseteq A^c$, whence by 1.1(2), $A^c \in \mathfrak{U}_x$, and by 1.3, A^c is open.

Conversely, let A^c be open, and let $x \in \overline{A}$. Suppose that $x \in A^c$, then by 1.3, $A^c \in \mathfrak{U}_x$, and by Ex. 1.33, $A^c \cap A \neq \varnothing$. This is clearly a contradiction, consequently $x \in A$, and $\overline{A} \subseteq A$. However, since for any set $A \subseteq \overline{A}$, we have $A = \overline{A}$, and A is closed. ∎

A simple application of this theorem, together with DeMorgan's rule, establishes the following:

1.15. Corollary. *In any topological space,*

(1) *The intersection of any number of closed sets is closed, and*
(2) *The union of any two (hence any finite number) of closed sets is closed.*

Proof. (1) Let A be an indexing set, and for each $\alpha \in A$ let C_α be closed. Further let $C = \bigcap_{\alpha \in A} C_\alpha$, then

$$C^c = \left(\bigcap_{\alpha \in A} C_\alpha \right)^c = \bigcup_{\alpha \in A} C_\alpha^c,$$

and since by 1.14 each C_α^c is open, so also is $\bigcup_{\alpha \in A} C_\alpha^c$ by 1.5. Thus C^c is open, whence again by 1.14, C is closed.

The proof of (2) is left as a trivial exercise. ∎

The following two definitions are needed in the exercises which follow. Some ambient topological space is assumed in each definition.

1.16. Definition. (1) *The **interior of A**, designated by $\mathring{A}$ (or $A°$) is the set of all x, such that $A \in \mathfrak{U}_x$.*

(2) *The **frontier of A**, designated by $\mathrm{Fr}\,(A)$ is the set $\overline{A} \cap \overline{A^c}$.*

1.17. Definition. *The set A is said to be **dense in the set B** if $\overline{A} \supseteq B$. If A is dense in the whole space, X, we say that A is everywhere dense, or sometimes, if there is no chance for misunderstanding, simply **dense**.*

EXERCISES

1.35. Show that $\mathring{A}$ is an open set, and is in fact the largest open set contained in A, in the sense that if B is open, $B \subseteq A$, then $B \subseteq \mathring{A}$.

1.36. Show that $\overline{A}$ is the smallest closed set which contains A, in the sense that if B is closed and $B \supseteq A$, then $B \supseteq \overline{A}$.

The preceding two exercises show that one might have defined the interior of A as the union of all open sets contained in A, and the closure of A as the intersection of all closed sets which contain A.

1.37. Give an example of a set A on the real line with the usual topology, such that from A 14 distinct sets may be formed, using only the two operations of complementation and closure. More generally, show that in any space the maximum number of distinct sets which may be formed from a given set using only complementation and closure is 14.

1.38. Show that

(a) $\operatorname{Fr}(A) = \operatorname{Fr}(A^c)$,

(b) $\operatorname{Fr}(\overline{A}) \subseteq \operatorname{Fr}(A)$ and $\operatorname{Fr}(\mathring{A}) \subseteq \operatorname{Fr}(A)$,

(c) $\operatorname{Fr}(A \cup B) \subseteq \operatorname{Fr}(A) \cup \operatorname{Fr}(B)$,

and in (b) and (c) give examples to show that proper containment can in fact occur.

1.39. If A and B are open show that

$$[A \cap \operatorname{Fr}(B)] \cup [B \cap \operatorname{Fr}(A)] \subseteq \operatorname{Fr}(A \cap B)$$
$$\subseteq [A \cap \operatorname{Fr}(B)] \cup [B \cap \operatorname{Fr}(A)] \cup [\operatorname{Fr}(A) \cap \operatorname{Fr}(B)].$$

1.40. If A is closed, show that $B \cap \operatorname{Fr}(A) \subseteq \operatorname{Fr}(A \cap B)$.

1.41. Show that

$$\operatorname{Fr}(A) = \{x \mid x \notin \mathring{A} \text{ and } x \notin \mathring{\overline{A^c}}\}$$
$$= \{x \mid x \text{ is in neither the interior of } A$$
$$\text{nor the interior of } A^c\}.$$

1.42. Prove 1.15(2).

1.43. Show that if X has the discrete topology, then $\{x\} \in \mathfrak{U}_x$. Further, if X has the discrete topology, $A \subseteq X$, what form do the sets $\overline{A}$, $\mathring{A}$, and $\operatorname{Fr}(A)$ have?

1.44. Let X be a set and let

$$\mathfrak{J} = \{\mathfrak{U}_x \mid x \in X\} \quad \text{and} \quad \mathfrak{J}' = \{\mathfrak{U}'_x \mid x \in X\}$$

be two topologies for X. Further let $\mathfrak{O}$ and $\mathfrak{O}'$ be the families of open sets determined by $\mathfrak{J}$ and $\mathfrak{J}'$ respectively.

(a) Show that $\mathfrak{J} = \mathfrak{J}'$ if and only if for each $x \in X$ and each $U \in \mathfrak{U}_x$ there exists $O' \in \mathfrak{O}'$ such that $x \in O' \subseteq U$, and for each $U' \in \mathfrak{U}'_x$ there exists $O \in \mathfrak{O}$, such that $x \in O \subseteq U'$.

(b) Show that $\mathfrak{J} = \mathfrak{J}'$ if and only if $\mathfrak{O} = \mathfrak{O}'$.

1.45. Let X be a set, and let there be assigned to each set $A \subseteq X$ another set $\overline{A} \subseteq X$, such that

I. For any $A, B \subseteq X$, $\overline{A \cup B} = \overline{A} \cup \overline{B}$.
II. If $A = \emptyset$ or A is a set consisting of a single point, then $\overline{A} = A$.
III. $\overline{\overline{A}} = \overline{A}$ for any set $A \subseteq X$.

Prove the following:

(1) $A \subseteq B$ implies $\overline{A} \subseteq \overline{B}$.
(2) $\overline{A \cap B} \subseteq \overline{A} \cap \overline{B}$.
(3) $\overline{A} - \overline{B} \subseteq \overline{A - B}$.
(4) $\overline{\bigcap_{\gamma \in C} A_\gamma} \subseteq \bigcap_{\gamma \in C} \overline{A_\gamma}$ where C is some indexing set.
(5) $\overline{\bigcup_{\gamma \in C} A_\gamma} \supseteq \bigcup_{\gamma \in C} \overline{A_\gamma}$.
(6) If A is finite, $\overline{A} = A$.
(7) $A \subseteq \overline{A}$ for every $A \subseteq X$.
(8) $\overline{X} = X$.
(9) Define $A \subseteq X$ to be closed provided that $A = \overline{A}$, then X is a topological space in which the family of open sets,

$$\mathfrak{O} = \{O \mid O = A^c, \text{ where } A \text{ is closed, i.e., } A = \overline{A}\}$$

and that with the topology so defined (using 1.6) the closure of a set $A \subseteq X$ is in fact $\overline{A}$.

1.46. Show by suitable example that the derived set, A', is not necessarily closed. [*Hint.* Let

$$X = \{x, y, z\}, \quad \mathfrak{O} = \{\emptyset, \{x, y\}, \{z\}, X\}, \quad A = \{x\}.]$$

§ 6 Subspaces

1.18. Definition. *Let X be a topological space, and let $Y \subseteq X$. For each $y \in Y$ define a neighborhood system $\mathfrak{V}_y$ in Y by $\mathfrak{V}_y = Y \cap \mathfrak{U}_y$, i.e., $V \in \mathfrak{V}_y$ if and only if $V = Y \cap U$ for some $U \in \mathfrak{U}_y$, where $\mathfrak{U}_y$ is the neighborhood system for y in X. The topology $\mathfrak{T}' = \{\mathfrak{V}_y \mid y \in Y\}$ is called the **topology induced on Y by the topology $\mathfrak{T}$ of X**, or more commonly the **relative topology of Y in X**, and $(Y, \mathfrak{T}')$ is called a **subspace** of X.*

It is important to note in this definition that a subset Y of a space X is not necessarily a subspace. Only if the topology of Y agrees with the induced (or relative) topology is Y called a subspace. The above definition is incomplete in that it is only asserted (without proof) that $\mathfrak{V}_y$ is a neighborhood system; by now the reader is inured to the fact that he is expected to prove that, in fact, $\mathfrak{V}_y$ satisfies the conditions of 1.1.

EXAMPLES (AND EXERCISES)

1.47. Let R be the real line, and E the real plane with the usual topologies, then R is a subspace of E, i.e., the usual topology in E induces the usual topology in $R = \{(x, y) \mid (x, y) \in E, y = 0\}$.

1.48. Let X and Y be nonempty sets, then the discrete topology in X induces the discrete topology in Y, and the trivial topology in X induces the trivial topology in Y.

1.49. Let $N = \{(x, y) \mid x, y \text{ real}, y \geqq 0\}$, i.e., N is the closed upper half of the real plane, and let

$$N° = \{(x, y) \mid x, y \text{ real}, y > 0\}.$$

For $(x, y) \in N$, define $\mathfrak{V}_{(x,y)} = \mathfrak{U}_{(x,y)} \cap N$ if $y > 0$, where $\mathfrak{U}_{(x,y)}$ is the neighborhood system for (x, y) in the usual topology for the real plane, and define

$$\mathfrak{V}_{(x,y)} = \{V \mid V \supseteq (U \cap N°) \cup \{(x, y)\}\}$$

for $U \in \mathfrak{U}_{(x,y)}$ if $y = 0$. Let

$$\mathfrak{T} = \{\mathfrak{V}_{(x,y)} \mid (x, y) \in N\},$$

with $\mathcal{U}_{(x,y)}$ as so defined, then $(N, \mathfrak{J})$ is a topological space. N is not a subspace of the real plane with the usual topology, nor is the real line with its usual topology a subspace of N. The real line with the discrete topology is a subspace of N, where we think of the real line here as the set $R = \{(x, y) \mid y = 0\}$.

If X is a topological space, and Y a subspace of X, there is a considerable distinction between $O \subseteq Y$ being open in Y (i.e., in the relative topology) and being open in X. For example, if X is the real line with the usual topology, and $Y = [0, 1]$, the closed unit interval, then the set $(\frac{1}{2}, 1]$ is an open subset of Y, since it is a neighborhood of each of its points, yet $(\frac{1}{2}, 1]$ is not open in X, since it is not a neighborhood of the point 1. We must, consequently, use some discretion when we talk about open sets in the context of spaces and subspaces, and must specify whether a given set is open in the whole space X, or in the subspace Y. A similar degree of discretion must of course be exercised in regard to closed sets, as well as other topological features, such as closure, limit point, and the like.

It would be gratifying if we had some criterion by which we could recognize the open sets of a subspace, and by a happy circumstance just such a criterion is given by the following:

1.19. Theorem. *Let X be a topological space, Y a subspace of X, then $O \subseteq Y$ is open in Y if and only if there exists a set O' open in X such that $O = O' \cap Y$.*

Proof. Let $\mathfrak{J} = \{\mathcal{U}_x \mid x \in X\}$ be the topology for X, and let

$$\mathfrak{J}' = \{\mathcal{V}_x \mid x \in Y\} = \{\mathcal{U}_x \cap Y \mid x \in Y\}$$

be the relative topology for Y. Suppose $O = O' \cap Y$, where O' is open in X, and let $x \in O$, then $x \in O'$, and $O' \in \mathcal{U}_x$ by 1.3. Thus

$$O = O' \cap Y \in \mathcal{U}_x \cap Y,$$

and O is open in Y.

Conversely, suppose O is open in Y, then $O \in \mathcal{U}_x \cap Y$ for each $x \in O$, that is $O = U_x \cap Y$ for some $U_x \in \mathcal{U}_x$ for each $x \in O$. Now for each $x \in O$, and each U_x, there exists O_x, open in X, such that $x \in O_x \subseteq U_x$, by 1.4. Define

$$O' = \bigcup_{x \in O} O_x,$$

then O' is open in X by 1.5. We now show that $O = O' \cap Y$. Let $y \in O \subseteq Y$, then $y \in O_y$, whence $y \in O'$, and $y \in O' \cap Y$. On the other hand, let $y \in O' \cap Y$, then $y \in O_x$ for some $x \in O$, and since $O_x \subseteq U_x$, $y \in U_x \cap Y = O$, whence $y \in O$. Thus $O = O' \cap Y$. ∎

EXERCISES

1.50. Prove that $\mathcal{U}_y$, as defined in 1.18, is a neighborhood system for $y \in Y$.

1.51. Let X be a topological space, and let Y be an open subset and a subspace of X. Show that O is open in Y if and only if O is open in X.

1.52. Show that if X is a topological space, Y a subspace of X, and if O is open in Y if and only if O is open in X, then Y is an open subset of X.

1.53. Show that if X is a topological space, Y a subspace of X, then $C \subseteq Y$ is closed in Y if and only if $C = C' \cap Y$ for some closed subset C' of X.

§ 7 Limits of Sequences; Hausdorff Spaces

1.20. Definition. *Let X be a topological space, let $x \in X$, and let $\{x_n \mid n = 1, 2, \ldots\}$ be a sequence of points in X. The sequence $\{x_n\}$ is said to **converge to** x, and x is said to be a **limit of the sequence** $\{x_n\}$ if and only if for each $U \in \mathcal{U}_x$, there exists an integer N such that $n \geq N$ implies $x_n \in U$.*

EXAMPLES (AND EXERCISES)

1.54. Let R be the real line with its usual topology, and let $\{x_n\}$ be a sequence in R, with x a limit of the sequence $\{x_n\}$. Show that $\lim x_n = x$ in the usual sense of analysis, i.e., for each $\epsilon > 0$, there is an integer N such that $n \geq N$ implies $|x_n - x| < \epsilon$.

1.55. Let X be a nonempty set with the discrete topology, let $\{x_n\}$ be a sequence in X, and let x be a limit of the sequence $\{x_n\}$, then there is an integer N such that for $n \geq N$, $x_n = x$.

1.56. Let E be the real plane, and define for $\epsilon > 0$,
$$S_\epsilon(x, y) = \{(u, v) \mid (u, v) \in E, |x - u| < \epsilon\}$$
and
$$\mathfrak{U}_{(x,y)} = \{U \mid U \supseteq S_\epsilon(x, y) \text{ for some } \epsilon > 0\}.$$

Verify that $\mathfrak{J} = \{\mathfrak{U}_{(x,y)} \mid (x, y) \in E\}$ is a topology for E. Let $\{(x_n, y_n)\}$ be a sequence in E, with the topology, $\mathfrak{J}$, and let (x_0, y_0) be a limit of the sequence $\{(x_n, y_n)\}$, then (x_0, z) for any z is also a limit of the sequence $\{(x_n, y_n)\}$, and observe thus that limits of sequences need not be unique.

1.57. *An alternate and somewhat more general approach to the notion of the limit of a sequence.* Let $\{A_n \mid n = 1, 2, \ldots\}$ be a sequence of sets in some topological space X. Define
$$\limsup_n A_n = \{y \mid U \cap A_n \neq \varnothing \text{ for each } U \in \mathfrak{U}_y \text{ and for infinitely many indices } n\};$$

$\limsup_n A_n$ is called the **limit superior of the sequence**
$$\{A_n \mid n = 1, 2, \ldots\}.$$
Also define
$$\liminf_n A_n = \{y \mid U \cap A_n \neq \varnothing \text{ for each } U \in \mathfrak{U}_y \text{ and for all but a finite number of the indices } n\};$$

$\liminf_n A_n$ is called the **limit inferior of the sequence**
$$\{A_n \mid n = 1, 2, \ldots\}.$$

If $\liminf_n A_n = \limsup_n A_n$, we define this common set to be $\lim_n A_n$. Prove the following:

(a) $\liminf_n A_n \subseteq \limsup_n A_n$.

(b) If $A_n = \{x_n\}$ for each n, and if $\lim_n A_n = \{x\}$ in the sense of the above definition, then $\lim_n x_n = x$ in the sense of Definition 1.20.

(c) Let $A_n = \{(x_n, y_n)\}, n = 1, 2, \ldots$ be the sequence of Ex. 1.56, then $\lim_n A_n = \{(x_0, z) \mid z \in R\}$ in the sense of the above definition.

For reasons that may not be apparent at this juncture, we now introduce a new kind of space, with a rather stronger structure. The motivation for the introduction of these spaces will appear after the definition and a few examples.

1.21. Definition. *Let $(X, \mathfrak{I})$ be a topological space, then $\mathfrak{I}$ is said to be a **Hausdorff** (or T_2) topology for X, provided that for each pair, x, y, with $x \neq y$, of points of X, there exist $U \in \mathfrak{U}_x$, $V \in \mathfrak{U}_y$ such that $U \cap V = \varnothing$. If this condition is satisfied we call X a **Hausdorff** (or T_2) **space**.*

EXAMPLES (AND EXERCISES)

1.58. (a) The real line with the usual topology is a Hausdorff space.
(b) The real plane with the usual topology is a Hausdorff space.

1.59. The real plane with the topology defined in Ex. 1.56 is *not* a Hausdorff space.

1.60. Let
$$X = \{f \mid f \text{ is a real-valued bounded function with domain } [0, 1]\},$$
define
$$S_\epsilon(f) = \{g \mid g \in X, |f(x) - g(x)| < \epsilon \text{ for all } x \in [0, 1]\},$$
and define
$$\mathfrak{U}_f = \{U \mid U \supseteq S_\epsilon(f) \text{ for some } \epsilon > 0\}.$$

The topology, $\mathfrak{I} = \{\mathfrak{U}_f \mid f \in X\}$, is then a Hausdorff topology for X. Observe that if $\{f_n\}$ is a sequence in X which converges to the limit f in the topology, $\mathfrak{I}$, then $\lim f_n(x) = f(x)$ uniformly on $[0, 1]$ in the ordinary sense of analysis, i.e., for each $\epsilon > 0$ there exists an integer N, such that $n \geq N$ implies

$|f_n(x) - f(x)| < \epsilon$ for all $x \in [0, 1]$.

As we have seen, in an arbitrary topological space, limits of sequences need not be unique. Since we are so used to having limits of sequences unique, it would be pleasant to have some property that would guarantee such uniqueness. This is precisely what motivates the introduction of Hausdorff spaces, for such spaces have the agreeable property that limits of sequences are unique, as in fact is shown by the following:

1.22. Theorem. *Let X be a Hausdorff space, and let $\{x_n\}$ be a sequence in X such that x is a limit of the sequence $\{x_n\}$, then x is **the** limit of $\{x_n\}$, i.e., the sequence $\{x_n\}$ has a unique limit if it has a limit.*

Proof. Suppose $\{x_n\}$ converges to both x and y, with $x \neq y$. Let $U \in \mathfrak{U}_x$, $V \in \mathfrak{U}_y$, such that $U \cap V = \varnothing$, by 1.21. Now since $\{x_n\}$ converges to x, there is an integer N_1 such that $n \geq N_1$ implies $x_n \in U$, and since $\{x_n\}$ converges to y, there is an integer N_2 such that $n \geq N_2$ implies $x_n \in V$. Select n sufficiently large so that $n > N_1$ and $n > N_2$, then simultaneously $x_n \in U$ and $x_n \in V$, whence $x_n \in U \cap V = \varnothing$. This is a palpable contradiction, and shows that $x = y$. ∎

It is entirely possible that a sequence, $\{x_n\}$, can converge to a point x, without x being a limit point of the set $\{x_n \mid n = 1, 2, \ldots\}$. For example, on the real line if $x_n = 1$ for all n, clearly $\{x_n\}$ converges to 1, yet 1 is not a limit point of the set $\{x_n \mid n = 1, 2, \ldots\}$. Also, it can occur that a sequence $\{x_n\}$ has a limit point, without having $\{x_n\}$ converge to that limit point, or for that matter to any point whatever. For example, on the real line, if $x_n = (-1)^n + 1/n$, then 1 is a limit point of $\{x_n\}$, yet $\{x_n\}$ does not converge to any point. These two examples make things look rather black, but we can be assured a bit by the following:

1.23. Theorem. *If X is a Hausdorff space, $\{x_n\}$ a sequence in X such that $\{x_n\}$ converges to some point $x \in X$, and if further y is a limit point of the set $\{x_n \mid n = 1, 2, \ldots\}$, then $x = y$.*

Proof. Suppose $x \neq y$. Then there exist $U \in \mathfrak{U}_x$, $V \in \mathfrak{U}_y$ such that $U \cap V = \varnothing$. Also, since $\{x_n\}$ converges to x, there exists an integer N, such that $n \geq N$ implies $x_n \in U$. Now by 1.21 there exists for each i, $1 \leq i \leq N$, $W_i \in \mathfrak{U}_{x_i}$, $V_i \in \mathfrak{U}_y$ such that $W_i \cap V_i = \varnothing$, unless $y = x_i$

for some i. Define $V'_i = V_i$ in case $x_i \neq y$, and $V'_i = V$ in case $x_i = y$, and define

$$V' = V \cap \left(\bigcap_{i=1}^{N} V'_i\right),$$

then $y \in V' \in \mathfrak{U}_y$ by 1.1(3). Now if z is such that $z \in V'$ and $z \neq y$, then $z \neq x_i$ for any i, since if $i \geq N$, $x_i \in U$, and $U \cap V' \subseteq U \cap V = \emptyset$, while if $i \leq N$, then either $x_i = y \neq z$, or $x_i \in W_i$, and $W_i \cap V' \subseteq W_i \cap V_i = \emptyset$. We have thus constructed a neighborhood V' of y, such that no point $z \neq y$ of the set $\{x_n \mid n = 1, 2, \ldots\}$ belongs to V'. This clearly contradicts the fact that y is a limit point of the set $\{x_n \mid n = 1, 2, \ldots\}$, hence $x = y$. ∎

§8 Comparison of Topologies

1.24. Definition. *Let X be a set and let $\mathfrak{J}_1$ and $\mathfrak{J}_2$ be two topologies for X; further for each $x \in X$ let $\mathfrak{U}_{i,x}$ be the family of neighborhoods of x in the topology $\mathfrak{J}_i$, $i = 1, 2$, then $\mathfrak{J}_1$ is **finer than** $\mathfrak{J}_2$ provided that $\mathfrak{U}_{2,x} \subseteq \mathfrak{U}_{1,x}$. In this case we write $\mathfrak{J}_2 \leq \mathfrak{J}_1$.*

In essence what the preceding definition says is that one topology is finer than another if each point has at least as many neighborhoods in the finer topology. The finest topology a space can have is clearly the discrete topology, whereas the coarsest (i.e., least fine) topology a space can have is the trivial topology. The finer the topology of a space, the richer the space is in neighborhoods; and so one might expect that it would also be richer in other structural features. This is, in fact, the case, as is shown in the following:

1.25. Theorem. *Let X be a set, and let $\mathfrak{J}_1$, $\mathfrak{J}_2$ be two topologies for X; further let $\mathfrak{O}_1$ and $\mathfrak{O}_2$ be the families of open sets generated by $\mathfrak{J}_1$ and $\mathfrak{J}_2$ respectively. Then $\mathfrak{J}_2 \leq \mathfrak{J}_1$ if and only if $\mathfrak{O}_2 \subseteq \mathfrak{O}_1$.*

Proof. Let $O \in \mathfrak{O}_2$. If $O = \emptyset$, then $O \in \mathfrak{O}_1$ by 1.5. If $O \neq \emptyset$, let $x \in O$, then by 1.3, $O \in \mathfrak{U}_{2,x}$, where $\mathfrak{U}_{i,x}$, $i = 1, 2$ has the meaning of 1.24. Since $\mathfrak{J}_2 \leq \mathfrak{J}_1$, we have by 1.24 that $\mathfrak{U}_{2,x} \subseteq \mathfrak{U}_{1,x}$, consequently $O \in \mathfrak{U}_{1,x}$, whence, since x was an arbitrary point of O, we have by 1.3, $O \in \mathfrak{O}_1$. Thus $\mathfrak{O}_2 \subseteq \mathfrak{O}_1$. The proof of the converse is left as an exercise. ∎

1.26. Corollary. Let X be a set, $\mathfrak{I}_1$ and $\mathfrak{I}_2$ two topologies for X such that $\mathfrak{I}_2 \leqq \mathfrak{I}_1$, and let $\mathfrak{C}_1$ and $\mathfrak{C}_2$ be the family of closed sets generated by $\mathfrak{I}_1$ and $\mathfrak{I}_2$ respectively, then $\mathfrak{C}_2 \subseteq \mathfrak{C}_1$.

Proof. Let $C \in \mathfrak{C}_2$, then $X - C \in \mathfrak{O}_2$ by 1.14. Thus by 1.25, $X - C \in \mathfrak{O}_1$, and again by 1.14, $C = X - (X - C) \in \mathfrak{C}_1$, and consequently $\mathfrak{C}_2 \subseteq \mathfrak{C}_1$. ∎

EXERCISES

1.61. Let X be a set, $\mathfrak{I}_1$ and $\mathfrak{I}_2$ two topologies for X, then $\mathfrak{I}_1 \leqq \mathfrak{I}_2$ if and only if for each $x \in X$ and each $U \in \mathfrak{U}_{1,x} \in \mathfrak{I}_1$ there is a $U' \in \mathfrak{U}_{2,x} \in \mathfrak{I}_2$ such that $U' \subseteq U$.

1.62. Let X be a set, $\mathfrak{I}_1$ and $\mathfrak{I}_2$ two topologies for X, then $\mathfrak{I}_1 = \mathfrak{I}_2$ if and only if both $\mathfrak{I}_1 \leqq \mathfrak{I}_2$ and $\mathfrak{I}_2 \leqq \mathfrak{I}_1$.

1.63. Prove the converse portion of 1.25.

§ 9 Bases, Countability Axioms, Separability

We have seen at least two ways to define a topology for a set X. We can simply define a neighborhood system for each point and take the family of all such neighborhood systems as the topology, or we can define a family of open sets, and define a neighborhood of a point as any set which contains an open set which contains the point. There is another way in which we can define a topology, namely by defining a base, as follows:

1.27. Definition. *A base* (or *basis*) *for a topology*, $\mathfrak{I}$, *of a space* X *is a set* $\mathfrak{B} \subseteq \mathfrak{O}$, *the family of open sets of* X, *such that for each* $x \in X$ *and each* $U \in \mathfrak{U}_x$, *there exists* $V \in \mathfrak{B}$ *such that* $x \in V \subseteq U$. *The sets of* $\mathfrak{B}$ *are called* **basic sets**.

Note that if $\mathfrak{B}$ is a basis for X and $Y \subseteq X$, then

$$\mathfrak{B} \cap Y = \{B' \mid B' = B \cap Y,\ B \in \mathfrak{B}\}$$

is a basis for the relative topology of Y. $\mathfrak{B} \cap Y$ is said to be the basis for Y induced by the basis $\mathfrak{B}$ for X. [Cf. Ex. 1.64(b) and Ex. 1.72.]

EXAMPLES (AND EXERCISES)

1.64. (a) Let R be the reals and let $\mathfrak{B} = \{(a, b) \mid a < b, a, b \in R\}$, i.e., the set of all open intervals, then $\mathfrak{B}$ is a basis for the usual topology of R.
(b) What is the basis for $[0, 1]$ in the relative topology induced by the basis in (a) above?

1.65. Let E be the real plane, and let $S_\epsilon(x, y)$ be an open ϵ-sphere about $(x, y) \in E$. Let
$$\mathfrak{B} = \{S_\epsilon(x, y) \mid \epsilon > 0, (x, y) \in E\},$$
then $\mathfrak{B}$ is a base for the usual topology of E.

1.66. Let X be a partially ordered set under "$\leq$," let
$$S_r(x) = \{y \mid y \geq x\},$$
and let $\mathfrak{B} = \{S_r(x) \mid x \in X\}$, then $\mathfrak{B}$ is a base for the right order topology of X.

1.67. Let X be the set of all real-valued integrable functions on $[0, 1]$, let
$$S_\epsilon(f) = \{g \mid g \in X, \int_0^1 |f - g| \, dx < \epsilon\},$$
and let $\mathfrak{B} = \{S_\epsilon(f) \mid f \in X, \epsilon > 0\}$, then $\mathfrak{B}$ is a base for the topology of X described in Ex. 1.4.

1.68. Let X be a nonempty set, and let $\mathfrak{B} = \{\{x\} \mid x \in X\}$, then $\mathfrak{B}$ is a basis for the discrete topology of X.

1.69. Let X be a nonempty set, and let $\mathfrak{B} = \{X\}$, then $\mathfrak{B}$ is a basis for the trivial topology of X.

The student should prove that the bases mentioned in the preceding examples are in fact bases for the topologies described.

1.28. Theorem. *Let X be a set, $\mathfrak{I}$ a topology for X, then $\mathfrak{B}$ is a base for $\mathfrak{I}$ if and only if the set*

$$\mathfrak{A} = \{\bigcup_{\alpha \in A} B_\alpha \mid A \text{ an arbitrary indexing set,}$$
$$B_\alpha \in \mathfrak{B} \text{ for each } \alpha\}$$

is in fact the family $\mathfrak{O}$ of open sets of X generated by $\mathfrak{T}$.

Proof. Suppose $\mathfrak{B}$ is a base for the topology $\mathfrak{T}$. Clearly, since each B_α is open, by 1.5(1) each element of $\mathfrak{A}$ is open, whence $\mathfrak{A} \subseteq \mathfrak{O}$. On the other hand let $O \in \mathfrak{O}$, and let $x \in O$, then by 1.3, $O \in \mathfrak{U}_x$, and by 1.27, there exists $B_x \in \mathfrak{B}$ such that $x \in B_x \subseteq O$. Clearly

$$O = \bigcup_{x \in O} B_x,$$

whence $O \in \mathfrak{A}$. Consequently $\mathfrak{O} \subseteq \mathfrak{A}$, and $\mathfrak{O} = \mathfrak{A}$.

Conversely, let $\mathfrak{B}$ have the required property, and let $x \in X$, $U \in \mathfrak{U}_x$, then by 1.4 there exists $O \in \mathfrak{O}$ such that $x \in O \subseteq U$. Now

$$O = \bigcup_{\alpha \in A} B_\alpha,$$

where $B_\alpha \in \mathfrak{B}$. Thus since $x \in O$, $x \in B_\alpha$ for some α, and $x \in B_\alpha \subseteq O \subseteq U$, whence by 1.27, $\mathfrak{B}$ is a base. ∎

EXERCISE

1.70. Let $X = \{0, 1, 2\}$, $A = \{0, 1\}$, $B = \{1, 2\}$, and let

$$\mathfrak{A} = \{X, A, B, \varnothing\},$$

then $\mathfrak{A}$ is not a basis for any topology for X. [*Hint.* Observe that any union of sets of $\mathfrak{A}$ is again in $\mathfrak{A}$, so that if $\mathfrak{A}$ were a basis for a topology $\mathfrak{T}$, $\mathfrak{A} = \mathfrak{O}$, the class of open sets of X generated by $\mathfrak{T}$. Show that $\mathfrak{A}$ cannot be a family of open sets.]

The viewpoint of Definition 1.27 is this: given a topology there exists a family, $\mathfrak{O}$, of open sets, and some subset of this is a basis. Now we might start from the other end, and ask: if we start with some family of subsets of a set X, can this family serve as a basis for some topology? The preceding exercise makes it clear that not every family of subsets of a given set, X, can serve as a basis for some topology or other. We might naturally ask, "Is there some property that distinguishes bases from other families of subsets of a given set?" It certainly would be pleasant if there were such a property, since it would enable us to distinguish bases from other families of subsets. It is, in fact, the case that such a property exists, as is shown in the following:

1.29. Theorem. *A family, $\mathcal{B}$, of subsets of a set X is a base for some topology, $\mathfrak{T}$, of X if and only if both*

(1) $X = \bigcup_{B_\alpha \in \mathcal{B}} B_\alpha$, *and*

(2) *For each $x \in X$ and each pair $U, V \in \mathcal{B}$, for which $x \in U$ and $x \in V$, there exists $W \in \mathcal{B}$ such that $x \in W \subseteq U \cap V$.*

Proof. Let $\mathcal{B}$ be a basis for some topology, $\mathfrak{T}$, for X, and let $x \in X$, then there exists $U \in \mathfrak{U}_x \subseteq \mathfrak{T}$, such that $x \in U$, and by 1.27 there exists $B_x \in \mathcal{B}$, such that $x \in B_x \subseteq U$. Clearly

$$X \subseteq \bigcup_{x \in X} B_x \subseteq \bigcup_{B_\alpha \in \mathcal{B}} B_\alpha \subseteq X,$$

hence $X = \bigcup_{B_\alpha \in \mathcal{B}} B_\alpha$, and condition (1) is met.

Let $U, V \in \mathcal{B}$, $x \in U$ and $x \in V$, define $U' = U \cap V$, then since by 1.27, U and V are both open, so also is U', whence $U' \in \mathfrak{U}_x$. Consequently, by 1.27, there exists $W \in \mathcal{B}$ such that $x \in W \subseteq U' = U \cap V$, and condition (2) is met.

Conversely, suppose $\mathcal{B}$ satisfies both conditions (1) and (2). Define

$$\mathcal{O} = \{\bigcup_{\alpha \in A} B_\alpha \mid A \text{ an arbitrary indexing set, } B_\alpha \in \mathcal{B}\}.$$

Now we wish to invoke 1.6 to show that $\mathcal{O}$ generates a topology $\mathfrak{T}$, so we need only show that $\mathcal{O}$ satisfies the hypotheses of 1.6. Since the union of members of $\mathcal{O}$ is again a union of sets of $\mathcal{B}$, such a union is again a member of $\mathcal{O}$, thus hypothesis (1) of 1.6 is satisfied. Condition (1) assures us that $X \in \mathcal{O}$, so that hypothesis (3) of 1.6 is satisfied. Let the indexing set $A = \emptyset$, then $\bigcup_{\alpha \in A} B_\alpha = \emptyset$, so that $\emptyset \in \mathcal{O}$, and hypothesis (4) of 1.6 is satisfied. Finally let $U, V \in \mathcal{O}$. If $U \cap V = \emptyset$, then since $\emptyset \in \mathcal{O}$, $U \cap V \in \mathcal{O}$. If $U \cap V \neq \emptyset$, then for each $x \in U \cap V$, select $B_x \in \mathcal{B}$, $B'_x \in \mathcal{B}$, such that

$$x \in B_x \subseteq U, \quad x \in B'_x \subseteq V.$$

By condition (2) there exists $W_x \in \mathcal{B}$, such that

$$x \in W_x \subseteq B_x \cap B'_x \subseteq U \cap V.$$

Finally let $W = \bigcup_{x \in U \cap V} W_x$, then clearly $U \cap V = W$, and $W \in \mathcal{O}$, since each $W_x \in \mathcal{B}$. Thus hypothesis (2) of 1.6 is satisfied, and $\mathcal{O}$ does indeed induce a topology in X. ∎

An important class of spaces are those for which there exists a basis with countably many elements—more formally:

1.30. Definition. *A space X, which has a base, $\mathcal{B}$, which is a countable family, i.e., $\mathcal{B} = \{B_i \mid i = 1, 2, \ldots\}$ is said to satisfy the **second axiom of countability** or sometimes more simply to have a **countable basis**.*

We sometimes also speak of spaces being second countable, meaning thereby that they have a countable basis, or satisfy the second axiom of countability.

EXAMPLES (AND EXERCISES)

1.71. (a) Let R be the reals with the usual topology, and let

$$\mathcal{B} = \{(a, b) \mid a, b \text{ rational}, a < b\},$$

then $\mathcal{B}$ is a countable basis for R.

(b) Let E be the real plane with the usual topology and let

$$\mathcal{B} = \{S_\epsilon(x, y) \mid x, y, \epsilon \text{ all rational}, \epsilon > 0\},$$

then $\mathcal{B}$ is a countable basis for E.

1.72. Show that any subspace of a space with a countable basis again has a countable basis. [*Hint.* If $Y \subseteq X$ and $\mathcal{B}$ is the basis for X, show first that $\mathcal{B} \cap Y = \{B \cap Y \mid B \in \mathcal{B}\}$ is a basis for Y.]

Closely related to second countable spaces are separable spaces, defined by:

1.31. Definition. *A space X is said to be **separable** provided it has a countable dense subset* (Cf. 1.17), *i.e., if there exists $A \subseteq X$, A countable such that $\overline{A} = X$.*

The relation between second countable spaces and separable spaces is given by:

1.32. Theorem. *Let X be a topological space with a countable basis, then X is separable.*

Proof. Let $\mathcal{B} = \{B_i \mid i = 1, 2, \ldots\}$ be a countable base for X, and define $A = \{x_i \mid x_i \in B_i, i = 1, 2, \ldots\}$, i.e., $A \cap B_i = \{x_i\}$. We prove now that $\overline{A} = X$. Let $x \in X$; if $x = x_i$ for some i, then $x \in A \subseteq \overline{A}$, so assume that $x \neq x_i$ for each i. Let $U \in \mathcal{U}_x$, then there exists $B_i \in \mathcal{B}$, such that $x \in B_i \subseteq U$. Now $x_i \in B_i \subseteq U$, and $x_i \neq x$, thus every neighborhood U of x contains a point of A distinct from x, whence $x \in \overline{A}$. Thus $X \subseteq \overline{A}$, and since $A \subseteq X$, $\overline{A} \subseteq \overline{X} = X$, we have $\overline{A} = X$, and A is the required countable dense subset. ∎

The converse of the preceding theorem is false, as is seen in the following:

EXAMPLES (AND EXERCISES)

1.73. Let R be the reals (or for that matter any uncountable set), and define the family, $\mathcal{O}$, of open sets by

$$\mathcal{O} = \{O \mid O = \emptyset, \text{ or } O = R - F, \text{ where } F \text{ is a finite set}\},$$

then $\mathcal{O}$ generates a topology for R by 1.6. Now let S be any infinite set, then $\overline{S} = R$; thus in particular if S is countable, we see that R is separable in this topology.

Suppose R has a countable base, $\mathcal{B}$, and let $x \in R$, then the set $R - \{x\}$ is open by definition of $\mathcal{O}$. Define

$$\mathcal{B}_x = \{B \mid B \in \mathcal{B}, x \in B\} \quad \text{and} \quad \mathcal{O}_x = \{O \mid O \in \mathcal{O}, x \in O\}.$$

It is easily shown that

$$\{x\} \subseteq \bigcap_{B \in \mathcal{B}_x} B \subseteq \bigcap_{O \in \mathcal{O}_x} O = \{x\},$$

whence $\{x\} = \bigcap_{B \in \mathcal{B}_x} B$. Now since each $B \in \mathcal{B}_x$ is open, $B^c = F$, a finite set, thus

$$R - \{x\} = R - \bigcap_{B \in \mathcal{B}_x} B = \bigcup_{B \in \mathcal{B}_x} B^c,$$

and this last set is a countable union (since $\mathcal{B}_x$, a subfamily of $\mathcal{B}$, is countable) of finite sets. Consequently $R - \{x\}$ is countable. This palpable contradiction shows that R, in the topology defined by $\mathcal{O}$, is not second countable.

1.74. Let J be the unit square in the plane, i.e.,

$$J = \{(x, y) \mid 0 \leq x \leq 1, 0 \leq y \leq 1\},$$

and let $\mathcal{J}$ be defined as follows: if

$$(x, y) \in \mathring{J} = \{(x, y) \mid 0 < x < 1, 0 < y < 1\}$$

define

$$\mathcal{U}_{(x,y)} = \{U \mid U \supseteq S_\epsilon(x, y) \cap J, \text{ for some } \epsilon > 0, \text{ where}$$
$$S_\epsilon(x, y) \text{ is an open } \epsilon\text{-sphere about } (x, y)\},$$

whereas if

$$(x, y) \in \text{Fr } (J) = \{(x, y) \mid x = 0, x = 1, y = 0 \text{ or } y = 1\}$$

define

$$\mathcal{U}_{(x,y)} = \{U \mid U \supseteq (S_\epsilon(x, y) \cap \mathring{J}) \cup \{(x, y)\}, \text{ for some } \epsilon > 0\}.$$

Then $(J, \mathcal{J})$ is separable, but has no countable basis.

The student may wonder why we have a second countability axiom, without a first. There is in fact a first; namely, a space is said to obey the **first axiom of countability, be first countable,** or have a **countable basis at each point,** provided that for each $x \in X$ there exists a countable family $\mathcal{B}_x \subseteq \mathcal{O}$ such that for each $U \in \mathcal{U}_x$ there exists $B \in \mathcal{B}_x$ such that $x \in B \subseteq U$. First countability is less demanding on the structure of a space than second countability, for clearly any second countable space is also first countable, while the following example shows the converse to be false.

EXAMPLES (AND EXERCISES)

1.75. Let X be an uncountable set with the discrete topology, let $x \in X$, then we may let $\mathcal{B}_x = \{B \mid B = \{x\}\}$, whence X is first countable, while clearly X is not second countable.

1.76. What is the relation between first countability and separability? That is to say, does either of these imply the other, or do there exist spaces which are separable but not first countable, and spaces which are first countable but not separable?

§ 10   Sub-bases, Product Spaces

We have seen how we can construct a topology for a space by beginning with either a neighborhood system (1.2) or a family of open sets (1.6) or a family of basic sets (1.29). Each of these in turn is a step removed from the original notion of neighborhood system. We can take another step backward without falling over the brink of the precipice and losing any chance of recovering a topology for a set. We take this last gingerly step as follows:

1.33. Definition. *Let X be a space, then $\mathcal{S}$ is called a **sub-base** for the topology, $\mathfrak{I}$, of X provided the family*

$$\left\{ B \mid B = \bigcap_{i=1}^{k} S_i,\, S_i \in \mathcal{S},\, k = 1, 2, \ldots \right\}$$

is a basis for $\mathfrak{I}$. That is to say, the family of sets obtained by taking finite intersections of members of $\mathcal{S}$ forms a basis for $\mathfrak{I}$.

Our freedom of choice as we move backward through the families $\mathcal{O}$ (1.6), $\mathcal{B}$ (1.29), $\mathcal{S}$ (1.33) becomes ever greater, as the hypotheses of 1.6, 1.29, and of the following theorem show.

1.34. Theorem. *Let X be a set, and let $\mathcal{S}$ be a family of subsets of X such that*

(1) $\mathcal{S} \neq \varnothing$,
(2) $X = \bigcup_{S \in \mathcal{S}} S$,

then $\mathcal{S}$ is a sub-base for some topology for X.

Proof. Let

$$\mathcal{B} = \left\{ B \mid B = \bigcap_{i=1}^{k} S_i,\, S_i \in \mathcal{S},\, k = 1, 2, \ldots \right\}.$$

Then clearly $\mathcal{B} \supseteq \mathcal{S}$, whence

$$X \supseteq \bigcup_{B \in \mathcal{B}} B \supseteq \bigcup_{S \in \mathcal{S}} S = X,$$

whence $X = \bigcup_{B \in \mathcal{B}} B$. Now let $x \in X$, and let $U, V \in \mathcal{B}$, with $x \in U$ and $x \in V$. Since U and V are finite intersections of elements of $\mathcal{S}$, so also is their intersection $W = U \cap V$, consequently $W \in \mathcal{B}$, and clearly

$$x \in W \subseteq U \cap V.$$

Thus the hypotheses of 1.29 are satisfied, and $\mathcal{S}$ generates $\mathcal{B}$ which in turn generates $\mathcal{T}$, a topology for X. ∎

The sub-base which gives rise to a topology according to the theorem and definition immediately preceding characterizes the topology it generates in a unique way. This is shown by the following:

EXAMPLE (AND EXERCISE)

1.77. Let $\mathcal{S}$ be a family of subsets of a set X, such that $\mathcal{S} \neq \emptyset$ and $\bigcup_{S \in \mathcal{S}} S = X$, let $\mathcal{T}$ be the topology generated by $\mathcal{S}$ according to 1.34, and let $\mathcal{O}$ be the family of open sets of X under $\mathcal{T}$. Then $\mathcal{T}$ is the coarsest topology for which $\mathcal{S} \subseteq \mathcal{O}$, in the sense that if $\mathcal{T}'$ is any other topology whose family of open sets is $\mathcal{O}'$, then if $\mathcal{S} \subseteq \mathcal{O}'$, it follows that $\mathcal{T} \leq \mathcal{T}'$.

It is rather a nuisance to have to write each time we want to specify some open set in a space, "$O \in \mathcal{O}$, the class of open sets of the space X," and we can avoid all this verbal fruit salad by agreeing that whenever we write "$\mathcal{O}$" we shall invariably mean the class of open sets of a space. Of course, similar remarks apply to the other families of subsets of a space which we have discussed, so let us make the following agreement: Whenever we write

- $\mathcal{U}_x$ we mean the family of neighborhoods of $x \in X$,
- $\mathcal{T}$ we mean the topology of X,
- $\mathcal{O}$ we mean the family of open sets of X,
- $\mathcal{C}$ we mean the family of closed sets of X,

$\mathcal{B}$ we mean some basis for the topology of X,
$\mathcal{S}$ we mean some sub-basis for the topology of X.

We shall be careful not to use these symbols to stand for anything other than the sets indicated in order, of course, to avoid any confusion.

One of the principal uses of sub-bases is in the introduction of a topology into Cartesian products of spaces. With the tools presently available to us it is difficult to describe the usual topology for a Cartesian product of infinitely many topological spaces; in fact, the very product itself is awkward to define until we have the notion of function. We thus restrict ourselves, for the time being, to Cartesian products with finitely many factors; and we put off until a later date, after we have had a chance to examine functions and continuous functions, the description of the standard method for defining and introducing topologies into Cartesian products of infinitely many topological spaces.

1.35. Definition. *Let X_i, $i = 1, 2, \ldots, n$, be topological spaces and let $\mathcal{O}_i$ be the associated families of open sets for each i. Let*

$$X = \underset{i=1}{\overset{n}{\times}} X_i$$

be their Cartesian product, and let

$$\mathcal{S} = \left\{ S \mid S = \underset{i=1}{\overset{n}{\times}} Y_i, \text{ where } Y_i = X_i \text{ for all } i \neq j, \text{ and } Y_j = U_j \in \mathcal{O}_j, j = 1, 2, \ldots, n \right\},$$

then $\mathcal{S}$ is a sub-base for the usual topology for X. The topology so defined is called the **product topology**.

The student should check (via 1.34) that the sub-base, $\mathcal{S}$, of 1.35 is, in fact, a sub-base for some topology. The points of $X = \underset{i=1}{\overset{n}{\times}} X_i$ are of the form $x = (x_1, x_2, \ldots, x_n)$ where $x_i \in X_i$ and the x_i are called coordinates of x. We may briefly describe the sub-base, $\mathcal{S}$, of 1.35 by saying that its members are simply collections of points of X in which all coordinates but one are free (i.e., may be any point in the associated space X_i) while one coordinate is restricted so as to lie in an open set (of X_j).

EXAMPLES (AND EXERCISES)

1.78. Let $E = R \times R$, R the real line, E consequently the real plane, and let $O \subseteq R$ be an open set of the reals in the usual topology, then $O = \bigcup_{\alpha \in A} I_\alpha$, where I_α is some open interval (Proof?) and we may assume that the I_α are disjoint. A sub-basic set for E in the product topology is a set of the form $R \times O$ (or $O \times R$) and is a collection of "rectangles" of infinite extent over the intervals of O. Basic sets of E in the product topology are certain collections of rectangles in E, while open sets of E are any collections of rectangles in E, where by a rectangle we mean a set of the form

$$\{(x, y) \mid a_1 < x < b_1, a_2 < y < b_2, a_1, b_1, a_2, b_2 \in R \text{ or } = \pm \infty\}.$$

It is important for the student to prove that the product topology for E is the same as the usual topology—essentially what must be proved is that for each $(x, y) \in E$ and each $\epsilon > 0$ there is a rectangle $R \subseteq E$ such that $(x, y) \in R \subseteq S_\epsilon(x, y)$ and for each rectangle $R \subseteq E$ such that $(x, y) \in R$ there is a $\delta > 0$ such that $S_\delta(x, y) \subseteq R$ (Cf. Ex. 1.7).

1.79. Let X_1, X_2 be topological spaces, $\mathfrak{I}_1$, $\mathfrak{I}_2$ their respective topologies, $\mathfrak{O}_1$, $\mathfrak{O}_2$ the associated families of open sets. Let $\mathfrak{B}$ be the basis for the product topology generated by $\mathcal{S}$ according to 1.35.
  (a) Prove that $\mathfrak{B} = \{U_1 \times U_2 \mid U_1 \in \mathfrak{O}_1, U_2 \in \mathfrak{O}_2\}$.
  (b) Prove that if X_1 and X_2 are Hausdorff spaces so also is $X_1 \times X_2$.

1.80. Let X be a topological space and let $Y \subseteq X$ be a subspace.
  (a) Prove that if X is second countable so also is Y.
  (b) Show by counterexample that X may be separable, yet Y may not be separable.

ADDITIONAL EXERCISES FOR CHAPTER 1

1.81. Let $\mathfrak{I}_1$, $\mathfrak{I}_2$ be two topologies for a set X, let $\mathfrak{B}_1$, $\mathfrak{B}_2$ be bases for $\mathfrak{I}_1$, $\mathfrak{I}_2$ respectively, then $\mathfrak{I}_1 = \mathfrak{I}_2$ if and only if for each $B_1 \in \mathfrak{B}_1$ and for each $x \in B_1$, there exists $B_2 \in \mathfrak{B}_2$ such that $x \in B_2 \subseteq B_1$, and also for each $B'_2 \in \mathfrak{B}_2$ and for each $x \in B'_2$ there exists $B'_1 \in \mathfrak{B}_1$ such that $x \in B'_1 \subseteq B'_2$.

1.82. Let X be any infinite set. Show that if we insist that every infinite subset of X be open then X must have the discrete topology.

1.83. Show that the set

$$\mathcal{S} = \{(a, b) \mid a = -\infty \text{ and } b \text{ real and finite,}$$
$$\text{or } a \text{ real and finite and } b = \infty\}$$

is a sub-basis for the usual topology of the reals.

1.84. In a Hausdorff space X, the set $\{x\}$ consisting of a single point is a closed set.

1.85. A lattice is a partially ordered set in which each pair of elements has a least upper bound and a greatest lower bound; more formally, if S is a set and "$\leq$" a relation in S such that
  (a) $x \leq x$ for all $x \in S$,
  (b) $x \leq y$ and $y \leq x$ imply $x = y$ for all $x, y \in S$,
  (c) $x \leq y$ and $y \leq z$ imply $x \leq z$ for all $x, y, z \in S$,
then S is said to be partially ordered under "$\leq$." A pair of elements, x, y, in S is said to have a least upper bound in S provided there exists $z \in S$ such that $x \leq z$ and $y \leq z$, and further if for any u, $x \leq u$ and $y \leq u$, then also $z \leq u$. A similar definition, with the inequalities reversed, yields the definition for greatest lower bound.

Now let $\mathfrak{F}$ be the family of all topologies on some set X, and let $\mathfrak{J}_1, \mathfrak{J}_2 \in \mathfrak{F}$. Further let $\mathfrak{O}_1$ and $\mathfrak{O}_2$ be the family of open sets generated by $\mathfrak{J}_1$ and $\mathfrak{J}_2$ respectively. In the set $\mathfrak{F}$ let "$\mathfrak{J}_2 \leq \mathfrak{J}_1$" mean that the topology $\mathfrak{J}_1$ is finer than the topology $\mathfrak{J}_2$. Further, define

$$\mathfrak{O}_* = \{O \mid O \in \mathfrak{O}_1 \text{ and } O \in \mathfrak{O}_2\} = \mathfrak{O}_1 \cap \mathfrak{O}_2;$$

and define $\mathcal{S} = \mathfrak{O}_1 \cup \mathfrak{O}_2$. Answer the following, giving proofs where appropriate:
  (i) Does $\mathfrak{O}_*$ satisfy the hypotheses of 1.6 and hence generate a topology, $\mathfrak{J}_*$?
  (ii) Is "$\leq$" a partial ordering for $\mathfrak{F}$?
  (iii) Is $\mathfrak{J}_*$ a greatest lower bound for $\mathfrak{J}_1$ and $\mathfrak{J}_2$?

(iv) Does $\mathcal{S}$ satisfy the hypotheses of 1.34, and is it hence a sub-base for some topology $\mathfrak{T}^*$?
(v) Is it so that $\mathfrak{T}_1 \leq \mathfrak{T}^*$ and $\mathfrak{T}_2 \leq \mathfrak{T}^*$?
(vi) Is $\mathfrak{T}^*$ a least upper bound for $\mathfrak{T}_1$ and $\mathfrak{T}_2$?
(vii) Is $\mathfrak{F}$ a lattice under "$\leq$"?

CHAPTER 2

CONTINUOUS FUNCTIONS (MAPPINGS) AND HOMEOMORPHISMS

§1 Functions

It might seem most natural at this point, when we have the general notion of a topology for a space pretty well under control, to try to classify topological spaces according to various properties we might conceive. We have already seen one such property, that of being a Hausdorff space. Natural though this path may seem, it is somewhat rocky, until we have a few further tools at our disposal; in particular we need the notion of continuous function (or mapping) on one topological space to another.

We take it for granted that the student is familiar with the notion of function, and list here simply some of the useful properties of functions. If the reader has never proved these for himself, as indeed he should sometime in his mathematical career, this is as good a time as any for him to verify them.

Let f be a function on X to Y; we write then simply $f : X \to Y$, and designate by $f(x)$, for $x \in X$, the value of f at the point (or place) x. We also call $f(x)$ the image of x under f. We say that two functions $f_1 : X \to Y$ and $f_2 : X \to Y$ are the same if $f_1(x) = f_2(x)$ for all $x \in X$, and write then $f_1 = f_2$. A **constant function** $f : X \to Y$ is a function such that for some fixed $y \in Y, f(x) = y$ for all $x \in X$. The **identity function on X**, $i : X \to X$, is the function i, such that $i(x) = x$ for all $x \in X$. The **image of $A \subseteq X$** under f is the set

$$f(A) = \{y \mid y \in Y, y = f(x) \text{ for some } x \in A\}.$$

The **restriction** of $f : X \to Y$ to $A \subseteq X$, written $f \mid A$ or f_A, is the function $f \mid A : A \to Y$ such that $(f \mid A)(x) = f(x)$ for each $x \in A$; under these circumstances f is called an **extension** of $f \mid A$ to X.

SEC. 2.1   CONTINUOUS FUNCTIONS AND HOMEOMORPHISMS

We observe the following properties for $f : X \to Y$, where $A \subseteq X$, $B \subseteq X$, $A_\gamma \subseteq X$ for each $\gamma \in G$, an indexing set:

1. $f(\varnothing) = \varnothing$.
2. $f(\{x\}) = \{f(x)\}$.
3. $A \subseteq B$ implies $f(A) \subseteq f(B)$.
4. $A \neq \varnothing$ implies $f(A) \neq \varnothing$.
5. $f(A \cup B) = f(A) \cup f(B)$ and more generally
$$f\left(\bigcup_{\gamma \in G} A_\gamma\right) = \bigcup_{\gamma \in G} f(A_\gamma).$$
6. $f(A \cap B) \subseteq f(A) \cap f(B)$ and more generally
$$f\left(\bigcap_{\gamma \in G} A_\gamma\right) \subseteq \bigcap_{\gamma \in G} f(A_\gamma).$$

The **inverse image** of a set $C \subseteq Y$ under f, designated by $f^{-1}(C)$, is the set $\{x \mid x \in X, f(x) \in C\}$, and we observe the following properties, where $f : X \to Y, C \subseteq Y, D \subseteq Y, A \subseteq X, C_\gamma \subseteq Y$ for each $\gamma \in G$, an indexing set:

1. $f(x) = y$ is equivalent to $x \in f^{-1}(y)$. [Note that we write $f^{-1}(y)$ in place of what is really correct, namely $f^{-1}(\{y\})$.]
2. $f^{-1}(\varnothing) = \varnothing$.
3. $C \subseteq D$ implies $f^{-1}(C) \subseteq f^{-1}(D)$.
4. $f^{-1}(C \cup D) = f^{-1}(C) \cup f^{-1}(D)$ and more generally
$$f^{-1}\left(\bigcup_{\gamma \in G} C_\gamma\right) = \bigcup_{\gamma \in G} f^{-1}(C_\gamma).$$
5. $f^{-1}(C \cap D) = f^{-1}(C) \cap f^{-1}(D)$ and more generally
$$f^{-1}\left(\bigcap_{\gamma \in G} C_\gamma\right) = \bigcap_{\gamma \in G} f^{-1}(C_\gamma).$$
6. $f^{-1}(C^c) = [f^{-1}(C)]^c$, where the superscript c indicates complementation with respect to X or Y as appropriate.
7. $A \subseteq f^{-1}[f(A)]$.
8. $f[f^{-1}(C)] \subseteq C$.

The function $f : X \to Y$ is said to be **onto** if $f(X) = Y$, and f is said to be **one-to-one** (or simply 1–1) if $x_1 \neq x_2$ implies $f(x_1) \neq f(x_2)$. We note the following properties, where $A \subseteq X, B \subseteq X$, and $C \subseteq Y$:

1. f is 1–1 if and only if $f(A \cap B) = f(A) \cap f(B)$ for all A and B.

2. f is onto if and only if $f^{-1}(C) \neq \emptyset$ for all $C \neq \emptyset$.
3. If f is 1–1 and onto, then f^{-1} is in fact a function on Y to X.
4. If f is 1–1 and onto, then $f(A^c) = [f(A)]^c$.
5. f is onto if and only if $f[f^{-1}(C)] = C$, for all C.
6. f is 1–1 if and only if $f^{-1}[f(A)] = A$, for all A.
7. f is 1–1 and onto if and only if for each $y \in Y$, $f^{-1}(y)$ is a single point of X.

If $f: X \to Y$ and $g: Y \to Z$ are functions, then $gf: X \to Z$ is the function defined by $gf(x) = g(f(x))$ for all $x \in X$. gf is sometimes written $g \circ f$. We note that $(gf)^{-1} = f^{-1}g^{-1}$, and that f is 1–1 and onto if and only if $f^{-1}f = i_X$ and $ff^{-1} = i_Y$, where i_X and i_Y are the identity functions on X to X and Y to Y respectively.

§ 2 Continuous Functions (Mappings)

We are now in a position to define a continuous function or mapping of one space into another. The motivating notion, intuitively at least, is the idea that points which are close together in X stay relatively close together in Y after f has acted on them. We have, via the notion of neighborhood, an idea of closeness; thus we make the following:

2.1. Definition. *Let* $(X, \Im)$ *and* $(Y, \Im')$ *be topological spaces, let* $f: X \to Y$ *be a function, and let* $x \in X$, *then f is* **continuous at x** *if for each* $V \in \mathfrak{U}_{f(x)} \in \Im'$ *there exists* $U \in \mathfrak{U}_x \in \Im$ *such that* $f(U) \subseteq V$. *The function f is said to be* **continuous** *if it is continuous at each point* $x \in X$.

EXAMPLES (AND EXERCISES)

2.1. Show that the identity function $i: X \to X$ is continuous.

2.2. Show that a constant function $c: X \to Y$ is continuous.

2.3. (a) Show that $f: X \to Y$ is continuous at $x \in X$ if and only if for each $V \in \mathfrak{U}_{f(x)}$, $f^{-1}(V) \in \mathfrak{U}_x$.
 (b) Show that if $f: X \to Y$ and $g: Y \to Z$ are continuous, so also is $gf: X \to Z$.

As we have just seen in the preceding exercises there are alternate ways to define continuity of a function at a point. Also there are alternate ways to define continuity of a function on a space X, the more important of which are given in the following:

2.2. Theorem. *Let X and Y be topological spaces, $f : X \to Y$ a function, then the following are logically equivalent statements:*

(1) *f is continuous.*
(2) *For every open set $O \subseteq Y$, $f^{-1}(O)$ is open in X.*
(3) *For every closed set $C \subseteq Y$, $f^{-1}(C)$ is closed in X.*
(4) *If $A \subseteq X$, then $f(\overline{A}) \subseteq \overline{f(A)}$.*

NOTE: The symbol "$\Rightarrow$" used in the following proof is read "implies," and we prove the theorem by showing that each statement implies the next and that the last statement implies the first.

Proof. (1) $\Rightarrow$ (2). Let $x \in f^{-1}(O)$, so that $f(x) \in O$. Since O is open it is a neighborhood of $f(x)$, thus by 2.1 there exists V a neighborhood of x such that $f(V) \subseteq O$, whence $V \subseteq f^{-1}(O)$ and $f^{-1}(O) \in \mathfrak{U}_x$. Thus by 1.3, $f^{-1}(O)$ is open.

(2) $\Rightarrow$ (3). Let $O = Y - C$, then O is open in Y and by (2), $f^{-1}(O)$ is open in X, and $X - f^{-1}(O) = [f^{-1}(O)]^c = f^{-1}(O^c) = f^{-1}(C)$ is closed.

(3) $\Rightarrow$ (4). We observe that $A \subseteq f^{-1}[f(A)] \subseteq f^{-1}[\overline{f(A)}]$, and since $\overline{f(A)}$ is closed, by (3), $f^{-1}[\overline{f(A)}]$ is also closed. Thus by 1.11(1), $\overline{A} \subseteq f^{-1}[\overline{f(A)}]$, whence $f(\overline{A}) \subseteq \overline{f(A)}$.

(4) $\Rightarrow$ (1). Let $x \in X$, and U be a neighborhood of $f(x)$ in Y. Select O open such that $f(x) \in O \subseteq U$, and let $V = [f^{-1}(O^c)]^c$, then since $f(x) \in O$, $f(x) \notin O^c$, $x \notin f^{-1}(O^c)$, and $x \in V$. We need only show $f^{-1}(O^c)$ closed, for then V will be open, hence a neighborhood of x, and

$$f(V) = f([f^{-1}(O^c)]^c) = f(f^{-1}[(O^c)^c]) = f(f^{-1}(O)) \subseteq O.$$

We now show $f^{-1}(O^c)$ closed. Since in any case $f^{-1}(O^c) \subseteq \overline{f^{-1}(O^c)}$, and since by (4)

$$f(\overline{f^{-1}(O^c)}) \subseteq \overline{f(f^{-1}(O^c))} \subseteq \overline{O^c} = O^c,$$

we have also

$$\overline{f^{-1}(O^c)} \subseteq f^{-1}(O^c),$$

thus $\overline{f^{-1}(O^c)} = f^{-1}(O^c)$, and $f^{-1}(O^c)$ is closed. ∎

The preceding theorem is important, since it is often necessary to show that a function is continuous, and this theorem gives us several methods for proving functions continuous. The name **mapping** is a useful term for continuous function and from now on we'll use it in that sense; that is, any function that we call a mapping will automatically be continuous.

Many of the useful and important properties of topological spaces are preserved under mappings. By this is meant the following: If X is a topological space which possesses some property P and if $f : X \to Y$ is a mapping into a topological space Y, then it is frequently the case (depending upon the property P as well as to some extent upon the structure of the spaces X and Y) that Y also possesses property P. Several examples of such properties occur in the following theorems.

2.3. Theorem. *Let X and Y be topological spaces, $f : X \to Y$ a 1-1 mapping, let $A \subseteq X$ and let $x \in A'$, then $f(x) \in f(A)'$.*

Proof. Let U be a neighborhood of $f(x)$, then by 2.1 there exists V, a neighborhood of x, such that $f(V) \subseteq U$. Since $x \in A'$, there exists $y \neq x, y \in V$ such that $y \in A$, whence $f(y) \in f(V) \subseteq U$ and $f(y) \in f(A)$. Further, since f is 1-1, $f(y) \neq f(x)$. Thus in each neighborhood U of $f(x)$ there is a point $f(y) \neq f(x)$ such that $f(y) \in f(A)$, whence $f(x) \in f(A)'$. ∎

2.4. Theorem. *Let X be a separable space, $f : X \to Y$ a mapping onto Y, then Y is separable.*

Proof. Let $\{x_n\}$ be a countable dense subset of X, then $\{f(x_n)\} \subseteq Y$ is again countable, and further we claim that $\{f(x_n)\}$ is dense in Y; for let $y \in Y$, and let U be any neighborhood of y, then there exists $x \in X$ such that $f(x) = y$ and there exists $V \in \mathfrak{U}_x$ such that $f(V) \subseteq U$. Since $\{x_n\}$ is dense in X there exists an integer i such that $x_i \in V$, whence $f(x_i) \in f(V) \subseteq U$, and $\{f(x_n)\}$ is dense in Y. ∎

NOTE: In the above theorem, since $f : X \to Y$ is a mapping onto, we say that Y is a **continuous image** of X.

§ 3 Homeomorphisms

Two spaces, one of which is a continuous image of the other, are, of course, related to some extent; however, there exists an even more intimate relation between spaces, which we now define.

2.5. Definition. *Let X and Y be topological spaces, and let $f : X \to Y$ be a one-to-one function from X onto Y (so that f^{-1} is also a function). Furthermore let both f and f^{-1} be continuous, then f is said to be a **homeomorphism**.*

We observe that the relation between spaces, $X \sim Y$, defined by $X \sim Y$ if and only if there exists a homeomorphism $f : X \to Y$, is an equivalence relation, for the identity function on X to X shows the relation to be reflexive. Also, if $f : X \to Y$ is a homeomorphism then $f^{-1} : Y \to X$ is one as well, and finally if $f : X \to Y$ and $g : Y \to Z$ are homeomorphisms, then so also is $gf : X \to Z$. We thus see that the relation "$\sim$" is symmetric and transitive as well. We may thus say that if a homeomorphism exists from X to Y (hence by the symmetry of the relation "$\sim$," between X and Y) then X and Y are homeomorphic; and we may also say that Y is a homeomorphic image of X.

Let us make a slight digression here. We are talking about an equivalence relation in the class of topological spaces. Now, as the reader knows, every equivalence relation in a set partitions the set into equivalence classes,

$$[X] = \{Y \mid Y \text{ is equivalent to } X\}.$$

In the case we are examining, $[X]$ is the set of all spaces homeomorphic to X. One of the problems of topology is to characterize the elements of $[X]$ for particularly well-known spaces X in terms other than their being simply homeomorphic to X. For example, if $X = [0, 1]$ with the relative topology inherited from the reals with the usual topology, is it possible to give certain characteristic properties of a space Y to guarantee that $Y \in [X]$ if and only if Y has these properties? The answer is "Yes," although we shall not explore far enough in this book to achieve this result. The interested reader can pursue these matters in more advanced texts on topology.* Now let us return to the main path of our discussion.

Many properties of spaces we shall eventually study are properties which are preserved by homeomorphisms. Specifically, if, when X has property P and X is homeomorphic to Y, then Y has property P also, we say that the property P is preserved by homeomorphisms, and we

---
* See, for example, J. G. Hocking and G. S. Young, *Topology* (Reading, Mass.: Addison-Wesley, 1961), p. 129.

call such a property a **topological property**. We remark at this juncture that any property that is preserved by continuous functions must also be preserved by homeomorphisms. Thus, among other things, 2.4 proves that separability is a topological property.

Before we continue with our study of homeomorphisms, it will be useful to have a few more terms to describe functions. These terms are given in the following:

2.6. Definition. *Let $f : X \to Y$ be a function, and let X and Y be topological spaces. If for each open set $O \subseteq X$, $f(O)$ is open in Y, f is said to be an **open function** (or sometimes an **interior function**). If for each closed set $C \subseteq X$, $f(C)$ is closed in Y, then f is said to be a **closed function**.*

EXAMPLES (AND EXERCISES)

2.4. Show that $f : X \to Y$ is a homeomorphism if and only if f is a one-to-one open mapping onto Y.

2.5. Show that $f : X \to Y$ is a homeomorphism if and only if f is a one-to-one closed mapping onto Y.

2.6. Let $f : X \to Y$ be a mapping, and let $Z \subseteq X$ be a subspace of X. Show that the restriction of f to Z, $f \mid Z : Z \to Y$ is also a mapping.

2.7. Let R be the real line with the usual topology, and let

$$K = \{(x, y) \mid x, y \in R, x^2 + y^2 = 1\}$$

be the unit circle with the relative topology inherited from the plane. Let $f : R \to K$ be defined by $f(x) = (\cos \pi x, \sin \pi x)$, i.e., f just wraps the real line round and round the unit circle. Show that f is not closed by showing the set

$$S = \{2n + 1/n \mid n = 1, 2, 3, \ldots\}$$

is a closed subset of R, but that $f(S)$ is not closed in K. Note, however, that f is open and continuous.

2.8. Let $I_1 = [-3, 3]$ with the relative topology inherited from the real line. Define $f : I_1 \to I_1$ by $f(x) = \frac{1}{6}(x^3 - 3x)$. By considering the set $S = \{x \mid x \in I_1, 0 < x \leq 3\}$ show that f is not open. It is clear that f is continuous and, as we shall see later, f is also closed.

Since the requirement that a function be a homeomorphism is more demanding than that it be merely continuous, we should expect that more properties are preserved under homeomorphisms than are preserved under mappings. Such is, in fact, the case, and the following theorems point this out.

2.7. Theorem. *Let X and Y be topological spaces, $f : X \to Y$ a homeomorphism, and let X be a Hausdorff space, then Y is a Hausdorff space.*

Proof. Let $y_1, y_2 \in Y$, with $y_1 \neq y_2$. Since f is one-to-one there exists a unique pair of points x_1 and x_2 such that $f(x_1) = y_1$ and $f(x_2) = y_2$. Since X is Hausdorff there exist U_1 and U_2 open and disjoint such that $x_1 \in U_1$, and $x_2 \in U_2$. Since f is open $f(U_1)$ and $f(U_2)$ are open in Y, and since f is one-to-one $f(U_1)$ and $f(U_2)$ are disjoint. Finally we note that $y_1 \in f(U_1)$ and $y_2 \in f(U_2)$, whence Y is Hausdorff. ∎

2.8. Theorem. *Let X and Y be topological spaces, $f : X \to Y$ a homeomorphism, and let X have a countable basis, then Y has a countable basis.*

Proof. Let $\mathcal{B}$ be a countable basis for X, and let $\mathcal{B}' = \{f(B) \mid B \in \mathcal{B}\}$, then since f is open $\mathcal{B}'$ is a family of open sets. Also, since $\mathcal{B}$ is countable, so also is $\mathcal{B}'$. Now let $y \in Y$, $U \in \mathcal{U}_y$, then there exist $x \in X$, $V \in \mathcal{U}_x$, $B \in \mathcal{B}$ such that $B \subseteq V$, and such that

$$y = f(x) \in f(B) \subseteq f(V) \subseteq U.$$

Letting $B' = f(B)$ we see that there exists $B' \in \mathcal{B}'$ such that $y \in B' \subseteq U$, and that consequently $\mathcal{B}'$ is a basis for Y. ∎

The preceding theorems might just as well be stated more concisely by saying that the property of being a Hausdorff space (or possession of a countable basis) is a topological property.

EXAMPLES (AND EXERCISES)

2.9. Let $f : X \to Y$ be a function, let X have the discrete topology, then f is continuous.

2.10. Let $f : X \to Y$ be a function, let Y have the trivial topology, then f is continuous.

2.11. Show by example that if $f : X \to Y$ is continuous and X is Hausdorff, Y need not be Hausdorff.

2.12. Show that $f : X \to Y$ is continuous if and only if for each $B \in \mathcal{B}$, a basis for Y, $f^{-1}(B)$ is open in X.

2.13. Let $f : X \to Y$ be continuous, let $\{x_n\}$ be a sequence in X such that $\lim\limits_n x_n = x$, show that $\lim\limits_n f(x_n) = f(x)$.

2.14. Let $f : X \to Y$ be a function; show that f is continuous if and only if for each $B \subseteq Y$, $\overline{f^{-1}(B)} \subseteq f^{-1}(\overline{B})$, and show by example that proper containment can occur.

2.15. Let X and Y be topological spaces, $X \times Y$ their product space, define the functions $P_X : X \times Y \to X$ and $P_Y : X \times Y \to Y$ by $P_X((x, y)) = x$ and $P_Y((x, y)) = y$ for each $(x, y) \in X \times Y$. Show that P_X and P_Y are continuous and open. P_X and P_Y are called projections of $X \times Y$ into the coordinate (or factor) spaces of $X \times Y$.

2.16. Let R be the reals with the usual topology and let X be the reals in which the topology is defined (via 1.6) by defining the family of open sets to be the complements of finite sets or the empty set, i.e.,

$$\mathcal{O} = \{O \mid O = \varnothing \text{ or } O^c \text{ is finite}\}.$$

Define $f : R \to X$ by $f(x) = x$, and show that f is one-to-one and continuous, but f^{-1} is not continuous.

2.17. (a) Let X and Y be topological spaces, Y a Hausdorff space, and let $f : X \to Y$ and $g : X \to Y$ be mappings. Let

$$E = \{x \mid x \in X, f(x) = g(x)\},$$

then E is closed.

(b) Further let $D \subseteq X$ be a dense subset of X, then if $D \subseteq E$, $f = g$ [i.e., $f(x) = g(x)$ for all $x \in X$].

(c) Let f be a real-valued continuous function of a real variable defined on the rational numbers. Show that there exists at most one function g defined for all real numbers with real values and continuous, such that $f(x) = g(x)$ for x rational.

2.18. Let $f : X \to Y$ be a function, show that f is continuous if and only if $f^{-1}(S)$ is open in X for each $S \in \mathcal{S}$, a sub-base for the topology of Y.

2.19. Let $f : X \to Y$ be a mapping. Let $A \subseteq X$, and let A have the relative topology, then $f \mid A : A \to Y$ is continuous.

2.20. Any two open intervals of the real line (with the usual relative topology) are homeomorphic. [A similar statement holds if "open" is replaced by "closed."]

2.21. Is the continuous image of a second countable space again second countable?

§4 Product Spaces

Exercise 2.15 immediately preceding gives us the device for extending the notion of product topology to Cartesian products with arbitrarily many factors. Let us begin with a product of two spaces X_1 and X_2, then a point (x_1, x_2) in $X_1 \times X_2$ can be thought of as a function on the set $\{1, 2\}$ whose value at $1 \in \{1, 2\}$ is x_1, and whose value at $2 \in \{1, 2\}$ is x_2. We can thus think of $X_1 \times X_2$ as the class of all functions $x(i)$, $i = 1, 2$ on the set $\{1, 2\}$ such that $x(1) \in X_1$ and $x(2) \in X_2$; however, instead of writing $x(1)$ and $x(2)$ we prefer to write x_1 and x_2. We now introduce a topology into the set $X_1 \times X_2$ by insisting that each projection P_{X_i}, $i = 1, 2$ be continuous and we can do this by insisting that $P_{X_i}^{-1}(U_i)$ be open where $U_i \subseteq X_i$, $i = 1, 2$ is an open set of X_i. Thus, in all, the product space $X_1 \times X_2$ can be thought of as the set of all functions x on the set $\{1, 2\}$ whose value at $i \in \{1, 2\}$ is x_i, with the topology generated by the sub-base

$$\{P_{X_i}^{-1}(U_i) \mid U_i \in \mathcal{O}_i, \mathcal{O}_i \text{ the family of open sets of } X_i, i = 1, 2\}.$$

The generalization to arbitrary indexing sets A, rather than the set $\{1, 2\}$, is now pretty open to us, and we make the following:

2.9. Definition. *Let A be an indexing set, X_α, $\alpha \in A$ a topological space, then the **product space** $\underset{\alpha \in A}{\times} X_\alpha$ is defined to be the set of all functions*

x on A such that x_α (the value of the function x at $\alpha \in A$) is an element of X_α, whose topology is generated by the sub-base

$$\{P_{X_\alpha}^{-1}(U_\alpha) \mid U_\alpha \in \mathcal{O}_\alpha, \text{ the family of open sets of } X_\alpha, \alpha \in A\},$$

where $P_{X_\alpha}(x) = x_\alpha$ is the projection of $\underset{\alpha \in A}{\times} X_\alpha$ into the "αth" coordinate space X_α.

Since we were motivated in making our definition by the requirement that each P_{X_α} be continuous, it is not surprising that at least half of the statement of the following theorem is correct.

2.10. Theorem. *Let* $X = \underset{\alpha \in A}{\times} X_\alpha$ *be the product space of the spaces* X_α, *where* A *is some indexing set. Let* $P_{X_\beta}(x) = x_\beta$ *be the projection of* X *onto* X_β, *then* P_{X_β} *is both open and continuous.*

Proof. First we prove continuity. Let $x \in X$ and let U be a neighborhood of $x_\beta \in X_\beta$. Select V open such that $x_\beta \in V \subseteq U$, then by 2.9, $P_{X_\alpha}^{-1}(V)$ is open in X, and $x \in P_{X_\beta}^{-1}(V)$, whence $P_{X_\beta}^{-1}(V)$ is a neighborhood of x. Finally $P_{X_\beta}(P_{X_\beta}^{-1}(V)) = V \subseteq U$, whence P_{X_β} is continuous.

We now prove P_{X_β} open. Let U be open in X, and let $x \in U$ (since if $U = \varnothing$, there is clearly nothing to prove), then U is a neighborhood of x. Since U is a union of basic neighborhoods, we may as well assume that U is itself a basic neighborhood, whence U is a finite intersection of sub-basic neighborhoods, i.e.,

$$U = \bigcap_{\alpha \in F} P_{X_\alpha}^{-1}(U_\alpha),$$

where $F \subseteq A$ is finite and U_α is open in X_α.

Now if $\beta \in F$, let $y \in U$, then since $y \in P_{X_\beta}^{-1}(U_\beta)$,

$$P_{X_\beta}(y) \in P_{X_\beta}(P_{X_\beta}^{-1}(U_\beta)) = U_\beta, \text{ and } P_{X_\beta}(U) = U_\beta,$$

thus if $\beta \in F$, P_{X_β} is open.

If $\beta \notin F$, we wish to show that $P_{X_\beta}(U) = X_\beta$. To this end let $z_\beta \in X_\beta$, and construct the point $z \in X$ such that $z_\alpha = x_\alpha \in U_\alpha$ for $\alpha \in F$, $z_\beta = z_\beta$ for $\beta \notin F$. Then since $x_\alpha \in U_\alpha$, $z \in P_{X_\alpha}^{-1}(U_\alpha)$ for each $\alpha \in F$, whence $z \in U$, and $P_{X_\beta}(z) = z_\beta$. Consequently $P_{X_\beta}(U) = X_\beta$, which is, of course, an open set in X_β. Thus if $\beta \notin F$, P_{X_β} is also open, whence in any case P_{X_β} is open. ∎

Now we ought to point out a few things here. First of all, it is not the case that P_{X_β} is closed. For example, in $R \times R$, the set

$$\{(x, y) \mid y = 1/(x - x^2),\ 0 < x < 1\}$$

is closed, yet the projection of this set on the x-axis is not closed, being the open unit interval. Secondly, it should be emphasized that if U_α is open in X_α, it does not necessarily follow that $\underset{\alpha \in A}{\times} U_\alpha$ is open in X, since in general it is not possible to write $\underset{\alpha \in A}{\times} U_\alpha$ as a union of *finite* intersections of sub-basic sets. Of course, if A happens to be finite, then $\underset{\alpha \in A}{\times} U_\alpha$ is open, but if A is infinite this need not be the case.

We now prove two theorems which are of general utility in many of the later theorems.

2.11. Theorem. *Let $f : Y \to \underset{\alpha \in A}{\times} X_\alpha$ be a function, then f is continuous if and only if the composite function $P_{X_\alpha} f : Y \to X_\alpha$ is continuous for each $\alpha \in A$.*

Proof. If f is continuous, then, since by 2.10 P_{X_α} is continuous, so also is $P_{X_\alpha} f$. On the other hand, let $P_{X_\alpha} f$ be continuous for each α, and let U_α be an open set in X_α, then

$$(P_{X_\alpha} f)^{-1}(U_\alpha) = f^{-1}[P_{X_\alpha}^{-1}(U_\alpha)]$$

is open in Y. However, $P_{X_\alpha}^{-1}(U_\alpha)$ is an arbitrary sub-basic set of the topology for X, hence by Ex. 2.18 f is continuous. ∎

2.12. Theorem. *Let X and Y be topological spaces, and let $f : X \to Y$ be a function. Define $F : X \to X \times Y$ by $F(x) = (x, f(x))$, then f is continuous if and only if F is a homeomorphism of X with the subspace $G = \{(a, f(a)) \mid a \in X\}$ of $X \times Y$.*

Proof. Let F be a homeomorphism of X with G, then by 2.11, $P_Y F$ is continuous. $P_Y F : X \to Y$ is defined by $(P_Y F)(x) = f(x)$, whence $P_Y F = f$, and f is continuous.

Conversely, suppose that f is continuous. Now we know from 2.10 that $P_X : X \times Y \to X$ is continuous, consequently $P_X \mid G$ is continuous;

however, observe that $(P_X \mid G)(x, f(x)) = x$, whence $P_X \mid G = F^{-1}$, and F^{-1} is continuous. We also remark that F is clearly one-to-one and onto G, thus we need only show that F is continuous.

Now $P_X F : X \to X$ is defined by $P_X F(x) = P_X[(x, f(x))] = x$, thus $P_X F$ is the identity function on X, hence is certainly continuous. Also $P_Y F : X \to Y$ is defined by $P_Y F(x) = P_Y[(x, f(x))] = f(x)$, whence $P_Y F = f$, and f is continuous by hypothesis. Therefore, since both $P_X F$ and $P_Y F$ are continuous, F is continuous by 2.11. This then establishes that F is a homeomorphism. ∎

The use of the letter G to describe the set mentioned in the previous theorem is not fortuitous, for the set G may be thought of as the "graph" of the function f. Thus 2.12 says in somewhat altered language that $f : X \to Y$ is continuous if and only if X is homeomorphic with the graph of f under the homeomorphism $F(x) = (x, f(x))$.

EXAMPLES (AND EXERCISES)

2.22. Let A be an indexing set, X_α a topological space for each $\alpha \in A$, and let $\underset{\alpha \in A}{\times} X_\alpha$ be the product space. Let $\beta \in A$ be fixed, and for each $\alpha \neq \beta$ let $x'_\alpha \in X_\alpha$ be some fixed point. Define

$$X'_\beta = \{x \mid x_\alpha = x'_\alpha \text{ for } \alpha \neq \beta, \text{ and } x_\beta \text{ arbitrary}\},$$

then $X'_\beta \subseteq \underset{\alpha \in A}{\times} X_\alpha$ and X'_β is homeomorphic to X_β.

2.23. Use the preceding exercise to show that the set of points

$$R' = \{(x, 1) \mid x \in R\}$$

is a subset of the plane and is homeomorphic to the reals.

CHAPTER 3

VARIOUS SPECIAL TYPES OF TOPOLOGICAL SPACES (VARIETIES OF COMPACTNESS)

In this chapter we shall study a variety of special restrictions one can put on a space so that a number of interesting and desirable properties result. Although we call these restrictions "special" in the sense that not every topological space need have them, they are nonetheless quite general in that many of the important and interesting topological spaces that we encounter frequently in analysis, algebraic topology, and elsewhere possess one or more of these properties.

§ 1 Compact Spaces

Before we begin the study of our first special property we need a preliminary notion, introduced in the following:

3.1. Definition. *Let X be a set, $Y \subseteq X$, and let*

$$\{D_\alpha \mid \alpha \in A, \text{ an indexing set}\}$$

*be a family of subsets of X, then $\{D_\alpha\}$ is called a **cover** or **covering** for Y provided $\bigcup_{\alpha \in A} D_\alpha \supseteq Y$.*

If all the sets, D_α, of a covering have some common property, P, we call the covering (or cover) a P-covering (or P-cover). For example, if each of the sets D_α is an open set, we call $\{D_\alpha\}$ an open cover. Further, if $\{E_\beta \mid \beta \in B\} \subseteq \{D_\alpha \mid \alpha \in A\}$ and $\bigcup_{\beta \in B} E_\beta \supseteq Y$ then $\{E_\beta \mid \beta \in B\}$ is called a subcovering of $\{D_\alpha \mid \alpha \in A\}$. We may also characterize the coverings of a set Y by properties of the indexing set A; for example, if

A is finite, we say that $\{D_\alpha \mid \alpha \in A\}$ is a finite covering, or if A is countable we say that $\{D_\alpha \mid \alpha \in A\}$ is a countable covering.

There is, of course, some ambiguity here, since the adjective preceding the word "covering" can refer either to some common property of the sets D_α or to some property of the indexing set A; however, the context will usually allow only one interpretation, and if more than one is possible, we will make it explicit whether the adjective refers to a property of A or to some property common to the D_α.

The first property of spaces we investigate is called compactness and is defined in the following:

3.2. Definition. *Let X be a topological space, then X is said to be* **compact** *provided each open cover of X contains a finite cover. (Here "open" refers to a property of the D_α, while "finite" refers to a property of the indexing set A.)*

The following picture of the notion may help. Suppose a large crowd of people (possibly infinite) is standing out in the rain, and suppose each of these people puts up his umbrella, then they will all stay dry. It is, of course, possible that they are all crowded so compactly together that not all, but merely a finite number of them need put up their umbrellas, and still they will all stay dry. We could then think of them as forming some sort of compact space. It is, of course, assumed in all this that the umbrellas are open.

EXAMPLES (AND EXERCISES)

3.1. (a) Any closed interval, $[a, b]$, $-\infty < a \leq b < \infty$ of the real line with the relative topology induced by the usual topology of the real line is a compact space. (This is one formulation of the Heine-Borel theorem.)

(b) More generally, any closed bounded set of real numbers is compact in the relative topology induced by the usual topology of the reals.

3.2. Let X be a partially ordered set with the right order topology, and let X have a smallest element (i.e., there is an $x \in X$ such that for all $y \in X$, $x \leq y$), then X is compact.

It should be noted at this point in the development that the notion of compactness for topological spaces is motivated by the Heine-Borel theorem, which asserts (in a version somewhat different from that given in the foregoing problems) for the real line, that if K is a closed bounded set of real numbers and $\{I_\alpha \mid \alpha \in A\}$ is a family of open intervals such that $\bigcup_{\alpha \in A} I_\alpha \supseteq K$, then a finite number of the I_α cover K.

A simple, but quite useful, consequence of 3.2 is the following:

3.3. Theorem. *Let X be a topological space, then X is compact if and only if for each family of closed sets $\{C_\alpha \mid \alpha \in A\}$ of X, $\bigcap_{\alpha \in A} C_\alpha = \emptyset$ implies that there exists $F \subseteq A$, F finite, such that $\bigcap_{\alpha \in F} C_\alpha = \emptyset$.*

Proof. Let X be compact, and let $\{C_\alpha \mid \alpha \in A\}$ be a family of closed sets with vacuous intersection. Define $O_\alpha = C_\alpha^c$, then O_α is open, and

$$\bigcup_{\alpha \in A} O_\alpha = \bigcup_{\alpha \in A} C_\alpha^c = \left[\bigcap_{\alpha \in A} C_\alpha\right]^c = \emptyset^c = X,$$

hence by 3.2 there exists $F \subseteq A$, F finite, such that $\bigcup_{\alpha \in F} O_\alpha = X$, whence

$$\emptyset = X^c = \left[\bigcup_{\alpha \in F} O_\alpha\right]^c = \bigcap_{\alpha \in F} O_\alpha^c = \bigcap_{\alpha \in F} C_\alpha.$$

The proof of the converse is left as an exercise. ∎

We wish to obtain some conditions equivalent to compactness beyond the one obtained in 3.3, and in order to do so it will be useful to have one further property of families of subsets of a space X, namely

3.4. Definition. *Let X be a set, $\{D_\alpha \mid \alpha \in A\}$ a family of subsets of X, then $\{D_\alpha\}$ is said to have the **finite intersection property** provided that for any finite, nonempty subset F of A, $\bigcap_{\alpha \in F} D_\alpha \neq \emptyset$.*

The following two equivalent formulations for compactness will be of considerable utility to us in what follows.

3.5. Theorem. *A space, X, is compact if and only if for any family $\{D_\alpha \mid \alpha \in A\}$ of closed sets with the finite intersection property $\bigcap_{\alpha \in A} D_\alpha \neq \emptyset$.*

Proof. Let X be compact, and let $\{D_\alpha \mid \alpha \in A\}$ be a family of closed sets with the finite intersection property. Suppose $\bigcap_{\alpha \in A} D_\alpha = \emptyset$, then by 3.3 there exists $F \subseteq A$, F finite, such that $\bigcap_{\alpha \in F} D_\alpha = \emptyset$, and this contradicts the fact that $\{D_\alpha\}$ has the finite intersection property. Thus $\bigcap_{\alpha \in A} D_\alpha \neq \emptyset$.

Conversely, let $\{O_\alpha \mid \alpha \in A\}$ be an open covering of X, and suppose that for each $F \subseteq A$, F finite, $\bigcup_{\alpha \in F} O_\alpha \not\supseteq X$, i.e., $\left(\bigcup_{\alpha \in F} O_\alpha\right)^c \neq \emptyset$. Define $D_\alpha = O_\alpha^c$, whence D_α is closed, and for any $F \subseteq A$, F finite,

$$\bigcap_{\alpha \in F} D_\alpha = \bigcap_{\alpha \in F} O_\alpha^c = \left(\bigcup_{\alpha \in F} O_\alpha\right)^c \neq \emptyset.$$

Thus $\{D_\alpha\}$ is a family of closed sets with the finite intersection property, and by hypothesis

$$\bigcap_{\alpha \in A} D_\alpha = \bigcap_{\alpha \in A} O_\alpha^c = \left(\bigcup_{\alpha \in A} O_\alpha\right)^c \neq \emptyset,$$

or $\bigcup_{\alpha \in A} O_\alpha \not\supseteq X$. This last statement, however, contradicts the fact that $\{O_\alpha\}$ is a cover for X, and this contradiction shows that there does exist a finite set $F \subseteq A$ such that $\bigcup_{\alpha \in F} O_\alpha = X$, whence X is compact. ∎

3.6. Theorem. *Let X be a topological space, then X is compact if and only if any family $\{D_\alpha \mid \alpha \in A\}$ of subsets of X with the finite intersection property has the further property that $\bigcap_{\alpha \in A} \overline{D_\alpha} \neq \emptyset$.*

Proof. Let X be compact, then since for each α, $\overline{D_\alpha} \supseteq D_\alpha$, it is clear that the family $\{\overline{D_\alpha} \mid \alpha \in A\}$ is a family of closed sets with the finite intersection property. Consequently by 3.5 $\bigcap_{\alpha \in A} \overline{D_\alpha} \neq \emptyset$.

Conversely, let any family $\{D_\alpha \mid \alpha \in A\}$ with the finite intersection property have the further property that the intersection of the closures of all the sets of the family is nonvacuous. Now let $\{C_\alpha \mid \alpha \in A\}$ be a family of closed sets with the finite intersection property, then $\overline{C_\alpha} = C_\alpha$, and

$$\bigcap_{\alpha \in A} C_\alpha = \bigcap_{\alpha \in A} \overline{C_\alpha} \neq \emptyset,$$

and again by 3.5, X is compact. ∎

EXERCISE

3.3. Prove the converse portion of Theorem 3.3.

We note now that compactness is preserved under mappings.

3.7. Theorem. *Let $f : X \to Y$ be a mapping onto and let X be compact, then Y is compact.*

Proof. Let $\{O_\alpha \mid \alpha \in A\}$ be an open covering of Y, then since f is continuous $f^{-1}(O_\alpha)$ is open in X for each α by 2.2(2), whence

$$\{f^{-1}(O_\alpha) \mid \alpha \in A\}$$

is an open covering of X. Thus, since X is compact there exists $F \subseteq A$, F finite, such that $\{f^{-1}(O_\alpha) \mid \alpha \in F\}$ is a cover of X. Finally, since f is onto,

$$\{f(f^{-1}(O_\alpha)) \mid \alpha \in F\} = \{O_\alpha \mid \alpha \in F\}$$

is a cover for Y, and therefore Y is compact. ∎

We now observe that as a simple consequence of the preceding theorem, compactness is a topological property, that is to say

3.8. Theorem. *Let X and Y be homeomorphic spaces, and let X be compact, then Y is compact.*

Let us now investigate briefly some relations between the properties of being compact and being closed.

3.9. Theorem. *Let X be a Hausdorff space, and let Y be a compact subspace of X, then Y is a closed subset of X.*

Proof. We shall show that Y^c is open. To this end, let $x \in Y^c$, then for each $y \in Y$ there exist neighborhoods O_{xy}, O_y which by 1.4 may be assumed to be open such that $x \in O_{xy}$, $y \in O_y$, and $O_{xy} \cap O_y = \varnothing$, since X is Hausdorff. Now $\{O_y \cap Y \mid y \in Y\}$ is an open cover for Y, thus since Y is compact in the relative topology, there exists a finite set F such that

$$Y = \bigcup_{y \in F} (O_y \cap Y).$$

Define $O = \bigcap_{y \in F} O_{xy}$, then $x \in O$, O is open and hence a neighborhood of x. Further

$$O \cap Y = O \cap \left(\bigcup_{y \in F} (O_y \cap Y) \right) \subseteq O \cap \left(\bigcup_{y \in F} O_y \right) = \emptyset,$$

for if there were a $z \in O \cap \left(\bigcup_{y \in F} O_y \right)$, we would have $z \in O_{y'}$ for some $y' \in F$, yet

$$z \in O = \bigcap_{y \in F} O_{xy} \subseteq O_{xy'},$$

whence $O_{y'} \cap O_{xy'} \neq \emptyset$, and this contradicts the choice of $O_{xy'}$ and $O_{y'}$.

Finally, since $O \cap Y = \emptyset$, $x \in O \subseteq Y^c$, whence Y^c is open, and from this it follows that Y is closed. ∎

3.10. Theorem. *Let X be a compact space, and let $Y \subseteq X$, Y closed, then Y is compact (in the relative topology).*

Proof. Let $\{O_\alpha \mid \alpha \in A\}$ be an open cover of Y in its relative topology, i.e., O_α is open in Y. Then by 1.18, $O_\alpha = Y \cap O'_\alpha$, where O'_α is open in X. Observe also that Y^c is open, thus

$$\{D_\alpha \mid D_\alpha = O'_\alpha \text{ or } D_\alpha = Y^c\}$$

is an open cover for X, since if $x \in X$, $x \in Y^c$ or

$$x \in Y \subseteq \bigcup_{\alpha \in A} O_\alpha \subseteq \bigcup_{\alpha \in A} O'_\alpha.$$

Now since X is compact, there exists $F \subseteq A$, F finite, such that

$$\left(\bigcup_{\alpha \in F} D_\alpha \right) \cup Y^c = X,$$

whence $\bigcup_{\alpha \in F} D_\alpha \supseteq Y$, where each of the D_α is of the form O'_α, consequently $Y \subseteq \bigcup_{\alpha \in F} D_\alpha = \bigcup_{\alpha \in F} O'_\alpha$, and finally

$$Y = \bigcup_{\alpha \in F} (O'_\alpha \cap Y) = \bigcup_{\alpha \in F} O_\alpha;$$

from this it follows that Y is compact. ∎

Up till this point the questions and answers (via theorems) have been relatively easy. We come now to a problem of an order of difficulty not encountered before. The question we want to answer is this: Suppose we have a family of spaces X_α, $\alpha \in A$, an indexing set, such that each X_α is compact, then is it the case that $\underset{\alpha \in A}{\times} X_\alpha$ is compact? The answer to this question is in the affirmative, and is given by a theorem due to Tychonoff, but we shall need a bit of machinery to get it. First we shall assume a lemma due to Zorn, which is in fact equivalent to the axiom of choice. The statement of this lemma is as follows:

3.11. Lemma (Zorn). *Let X be a partially ordered set, and let each simply ordered subset of X have an upper bound in X, then X has a maximal element.* [*That is, if for each $Y \subseteq X$, such that Y is simply ordered, there exists $z \in X$ such that $y \in Y$ implies $y \leqq z$, then there exists $m \in X$ such that for any $x \in X$, either x and m are incomparable (i.e., neither $x \leqq m$ nor $m \leqq x$ is true) or $x \leqq m$.*]

We make no attempt to derive this lemma from the axiom of choice. The interested student can find such derivations in any of a number of texts on set theory.

In preparation for proving Tychonoff's theorem we prove first a lemma which is more set-theoretical than topological, but which will considerably shorten the proof of Tychonoff's theorem. We have in mind the use of the finite intersection property for the proof that $X = \underset{\alpha \in A}{\times} X_\alpha$ is compact where each X_α is compact, and that is why the following lemma concentrates on sets with the finite intersection property.

3.12. Lemma. *Let X be a set, $\mathfrak{F} = \{F\}$ a family of subsets of X such that $\mathfrak{F}$ has the finite intersection property, then there exists a family $\mathfrak{M} = \{M\}$ of subsets of X with the following properties:*

(1) $\mathfrak{M}$ *has the finite intersection property.*
(2) $\mathfrak{M} \supseteq \mathfrak{F}$.
(3) $\mathfrak{M}$ *is maximal with respect to properties 1 and 2 (i.e., if $\mathfrak{M} \subseteq \mathfrak{N}$, $\mathfrak{N} \neq \mathfrak{M}$, $\mathfrak{N} \supseteq \mathfrak{F}$, then $\mathfrak{N}$ does not have the finite intersection property).*

(4) *Any finite intersection of elements of $\mathfrak{M}$ is again an element of $\mathfrak{M}$.*

Proof. Define

$$\mathfrak{B} = \{\mathcal{G} \mid \mathcal{G} = \{G\}, G \subseteq X, \mathcal{G} \supseteq \mathfrak{F}, \mathcal{G} \text{ has the finite intersection property}\},$$

that is, $\mathfrak{B}$ is the class of all families of subsets of X which contain $\mathfrak{F}$ and also have the finite intersection property. We introduce a partial ordering into $\mathfrak{B}$ as follows: let $\mathfrak{G}_1, \mathfrak{G}_2 \in \mathfrak{B}$, then $\mathfrak{G}_1 \leq \mathfrak{G}_2$ means $\mathfrak{G}_1 \subseteq \mathfrak{G}_2$.

We now show that the hypotheses of Zorn's lemma are satisfied in $\mathfrak{B}$ with the partial ordering "$\leq$." To this end, let $\mathfrak{G}_\alpha$, $\alpha \in A$, where A is some indexing set, be a simply ordered subset of $\mathfrak{B}$. Define $\mathfrak{G} = \bigcup_{\alpha \in A} \mathfrak{G}_\alpha$, i.e., $\mathfrak{G}$ contains as elements any set which occurs in any $\mathfrak{G}_\alpha$. We now verify that $\mathfrak{G} \in \mathfrak{B}$. First of all, since each $\mathfrak{G}_\alpha \supseteq \mathfrak{F}$, it follows that $\mathfrak{G} \supseteq \mathfrak{F}$; also, since each $B \in \mathfrak{G}$ is obtained as some element of some $\mathfrak{G}_\alpha$, each such $B \subseteq X$. Finally we must see that $\mathfrak{G}$ has the finite intersection property. Thus let $B_1, B_2, \ldots, B_k \in \mathfrak{G}$. Now $B_i \in \mathfrak{G}_{\alpha_i}$ for each i, by definition of $\mathfrak{G}$, and since the $\mathfrak{G}_\alpha$ are simply ordered, there exists a j, $1 \leq j \leq k$, such that $\mathfrak{G}_{\alpha_i} \leq \mathfrak{G}_{\alpha_j}$ for all i, $1 \leq i \leq k$. This means that $\mathfrak{G}_{\alpha_i} \subseteq \mathfrak{G}_{\alpha_j}$ for all i, $1 \leq i \leq k$. Now $\mathfrak{G}_{\alpha_j}$ has the finite intersection property, and each $B_i \in \mathfrak{G}_{\alpha_j}$, $1 \leq i \leq k$, thus

$$\bigcap_{i=1}^{k} B_i \neq \varnothing,$$

whence $\mathfrak{G}$ has the finite intersection property. Finally it is clear from the definition of $\mathfrak{G}$ that $\mathfrak{G} \supseteq \mathfrak{G}_\alpha$ for each α, hence $\mathfrak{G}_\alpha \leq \mathfrak{G}$, and consequently each simply ordered subset of $\mathfrak{B}$ has an upper bound in $\mathfrak{B}$.

We are now in a position to apply Zorn's lemma to $\mathfrak{B}$. We have seen that $\mathfrak{B}$ with the partial ordering introduced into $\mathfrak{B}$ satisfies the hypotheses of Zorn's lemma, thus let $\mathfrak{M}$ be a maximal element of $\mathfrak{B}$, whose existence is guaranteed by 3.11; then, since $\mathfrak{M} \in \mathfrak{B}$, we have

(1) $\mathfrak{M}$ has the finite intersection property.

(2) $\mathfrak{M} \supseteq \mathfrak{F}$.

(3) If $\mathfrak{N} \supseteq \mathfrak{M}$, then $\mathfrak{N} \geq \mathfrak{M}$, and if $\mathfrak{N} \supseteq \mathfrak{F}$ and $\mathfrak{N}$ has the finite intersection property, then $\mathfrak{N} \in \mathfrak{B}$, consequently if $\mathfrak{N} \neq \mathfrak{M}$, $\mathfrak{N} > \mathfrak{M}$, and this would contradict the maximality of $\mathfrak{M}$. We see thus that $\mathfrak{M}$ is maximal with respect to the finite intersection property.

(4) Let $M_1, M_2, \ldots, M_k \in \mathfrak{M}$, and let $M_0 = \bigcap_{i=1}^{k} M_i \neq \emptyset$, since $\mathfrak{M}$ has the finite intersection property. Now let $\mathfrak{M}' = \mathfrak{M} \cup \{M_0\}$, and let $N_1, N_2, \ldots, N_p \in \mathfrak{M}'$. If $N_i, 1 \leq i \leq p$ are all elements of $\mathfrak{M}$, then

$$\bigcap_{i=1}^{p} N_i \neq \emptyset$$

since $\mathfrak{M}$ has the finite intersection property, and if for some j, $N_j = M_0$, then

$$\bigcap_{i=1}^{p} N_i = \left(\bigcap_{i \neq j} N_i\right) \cap \left(\bigcap_{i=1}^{k} M_i\right) \neq \emptyset$$

where $N_i \in \mathfrak{M}, i \neq j$, whence the intersection is nonempty since $\mathfrak{M}$ has the finite intersection property. Also clearly $\mathfrak{M}' \supseteq \mathfrak{M}$ and $\mathfrak{M}' \supseteq \mathfrak{F}$, thus $\mathfrak{M} \leq \mathfrak{M}'$, and since $\mathfrak{M}$ is maximal $\mathfrak{M} = \mathfrak{M}'$, i.e., $M_0 \in \mathfrak{M}$, whence any finite intersection of elements of $\mathfrak{M}$ is again an element of $\mathfrak{M}$. ∎

We now prove the Tychonoff theorem.

3.13. Theorem. *Let X_α be a compact space for each $\alpha \in A$, an indexing set, then $X = \underset{\alpha \in A}{\times} X_\alpha$, with the product topology, is also compact.*

Proof. Let $\mathfrak{F} = \{F\}$ be any family of subsets of X, such that $\mathfrak{F}$ has the finite intersection property. Let $\mathfrak{M}$ be the family whose existence is guaranteed by 3.12 such that

(1) $\mathfrak{M}$ has the finite intersection property.
(2) $\mathfrak{M} \supseteq \mathfrak{F}$.
(3) $\mathfrak{M}$ is maximal with respect to the finite intersection property.
(4) Finite intersections of elements of $\mathfrak{M}$ are again in $\mathfrak{M}$.

Define $\mathfrak{M}_\alpha = P_{X_\alpha} \mathfrak{M}$, i.e., $M_\alpha \in \mathfrak{M}_\alpha$ if and only if $M_\alpha = P_{X_\alpha} M$ for some $M \in \mathfrak{M}$. Now the finite intersection property is clearly inherited from $\mathfrak{M}$ to $\mathfrak{M}_\alpha$, and since X_α is compact and $\mathfrak{M}_\alpha$ is a family of subsets of X_α, we have by 3.6 that

$$\bigcap_{M_\alpha \in \mathfrak{M}_\alpha} \overline{M_\alpha} \neq \emptyset.$$

Now for each α select $x_\alpha \in \bigcap_{M_\alpha \in \mathfrak{M}_\alpha} \overline{M_\alpha}$, and define $x = (x_\alpha)$, i.e., x is the point whose αth coordinate is x_α.

We now prove that $x \in \bigcap_{M \in \mathfrak{M}} \overline{M}$. Let U be a neighborhood of x, then there exists a basic set B such that $x \in B \subseteq U$. Now

$$B = \bigcap_{i=1}^{k} S_i,$$

where S_i are sub-basic sets for the product topology, i.e.,

$$S_i = \underset{\alpha \in A}{\times} Y_\alpha, \qquad Y_\alpha = X_\alpha$$

with at most one exception, α', and $Y_{\alpha'} = O_{\alpha'}$, an open set of $X_{\alpha'}$. Now since $x \in B \subseteq S_i$ for each i, $x_\alpha \in P_{X_\alpha} S_i = Y_\alpha$, and Y_α is an open set of X_α, whether $Y_\alpha = X_\alpha$ or O_α. Then $x_\alpha \in \overline{M_\alpha}$ for each $M_\alpha \in \mathfrak{M}_\alpha$, hence $Y_\alpha \cap M_\alpha \neq \emptyset$.

Thus $M \cap S_i \neq \emptyset$ for each i and for each $M \in \mathfrak{M}$, and consequently the adjunction of S_i to $\mathfrak{M}$ will not destroy the finite intersection property of $\mathfrak{M}$. Thus since $\mathfrak{M}$ is maximal with respect to the finite intersection property, $S_i \in \mathfrak{M}$, for each i. Further since finite intersections of elements of $\mathfrak{M}$ still belong to $\mathfrak{M}$,

$$B = \bigcap_{i=1}^{k} S_i \in \mathfrak{M}.$$

Let $M \in \mathfrak{M}$, then since $B \in \mathfrak{M}$, and $\mathfrak{M}$ has the finite intersection property, $M \cap B \neq \emptyset$, whence $x \in \overline{M}$ for each $M \in \mathfrak{M}$, and $x \in \bigcap_{M \in \mathfrak{M}} \overline{M}$. Thus

$$\bigcap_{F \in \mathfrak{F}} \overline{F} \supseteq \bigcap_{M \in \mathfrak{M}} \overline{M},$$

since $\mathfrak{F} \subseteq \mathfrak{M}$ and again by 3.6, X is compact. ∎

EXAMPLES (AND EXERCISES)

3.4. Any closed bounded set of the real plane with the usual topology (i.e., that topology induced by the usual topology of the plane) is compact.

3.5. Show that if X is compact, Y Hausdorff, $f : X \to Y$ a mapping, then $f(X)$ is a closed subset of Y.

3.6. Verify that the partial ordering introduced into $\mathfrak{B}$ in 3.12 is indeed a partial ordering.

3.7. Let $X = \underset{\alpha \in A}{\times} X_\alpha$ be a compact space in the product topology. Prove that X_α is compact for each α. [*Hint.* Use 3.7.]

§ 2 Separation Axioms

We now take a moment off from considering compact spaces to consider the extent to which points or closed sets can be separated from one another by means of open sets. We introduce a sequence of axioms, called separation axioms, into the definition of a space as follows:

3.14. Definition. *Let X be a topological space, then X is said to be a T_i space, provided it satisfies Axiom T_i, $i = 0, 1, 2, 3, 4$, where the axioms are as follows:*

Axiom T_0: *For each x and $y \in X$, $x \neq y$, either there exists $U \in \mathfrak{U}_x$ such that $y \notin U$, or there exists $V \in \mathfrak{U}_y$ such that $x \notin V$.*

Axiom T_1: *For each x and $y \in X$, $x \neq y$, there exist $U \in \mathfrak{U}_x$ and $V \in \mathfrak{U}_y$ such that $y \notin U$ and $x \notin V$.*

Axiom T_2: *For each x and $y \in X$, $x \neq y$ there exist $U \in \mathfrak{U}_x$ and $V \in \mathfrak{U}_y$ such that $U \cap V = \varnothing$.*

Axiom T_3: *For each $x \in X$ and each closed set $C \subseteq X$, $x \notin C$, there exist $U \in \mathfrak{U}_x$ and $O \in \mathfrak{O}$, the class of open sets of X, such that $C \subseteq O$ and $O \cap U = \varnothing$.*

Axiom T_4: *For each pair of closed disjoint sets, $C, D \subseteq X$, there exists a pair $O_1, O_2 \in \mathfrak{O}$, such that $C \subseteq O_1$, $D \subseteq O_2$, and $O_1 \cap O_2 = \varnothing$.*

Before we make any remarks about this definition, we prove a simple theorem.

3.15. Theorem. *A space X is a T_1 space if and only if each point is closed.*

Proof. Let $x \in X$, then for each $y \in X$, $y \neq x$, select $U \in \mathfrak{U}_y$ such that $x \notin U$. Then $X - \{x\} \supseteq U$, whence $X - \{x\} \in \mathfrak{U}_y$ for each $y \in X - \{x\}$, whence by 1.3, $X - \{x\}$ is open, and $\{x\}$ is closed.

Conversely, let $x \in X$, $y \in X$, and $x \neq y$, then since $\{x\}$ is closed, $X - \{x\}$ is open, and since $y \in X - \{x\}$, $X - \{x\} \in \mathfrak{U}_y$, and by an identical argument $X - \{y\} \in \mathfrak{U}_x$, whence X is a T_1 space. ∎

The language in which the above theorem is couched is open to criticism, since points cannot really be closed. It is rather the set $\{x\}$ that is closed. We will feel free, however, to say "points are closed" since this is much more expressive and so much simpler to say, and have it understood that what we really mean is that sets of the form $\{x\}$ are closed sets.

Now let us examine 3.14 more closely. First of all, we observe that Axiom T_2 is just the statement that a space is a Hausdorff space, so that the class of Hausdorff spaces and the class of T_2 spaces are identical. We leave the rest of the discussion to the following:

EXAMPLES (AND EXERCISES)

3.8. Let X be a partially ordered set with the right order topology, then X is a T_0 space, but not necessarily a T_1 space.

3.9. Let X be any infinite set, and define

$$\mathfrak{O} = \{O \mid O = \varnothing \text{ or } O = F^c, \text{ where } F \text{ is finite}\},$$

then X with the topology induced by $\mathfrak{O}$ via 1.6 is a T_1 space, but not a T_2 space.

3.10. Let X have the trivial topology, then X is a T_3 space, but not necessarily a T_2 space, since a single point need not be a closed set.

3.11. Let X be a T_1 space, x a limit point of $A \subseteq X$, then for each $U \in \mathfrak{U}_x$, $U \cap A$ is an infinite set.

3.12. Let X be a T_1 space, x a limit point of $A \subseteq X$, F a finite set in X, then x is a limit point of $A - F$.

3.13. Let X be a T_1 space, F a finite set in X, then F has no limit points.

As is shown in the preceding exercises, it is not the case that each axiom T_i is stronger than the preceding axiom T_{i-1}, for we have seen that there exist spaces which are T_3 but not necessarily T_2. It would be pleasant if we had a sequence of successively stronger axioms, such that each one implied the preceding one. To the end of obtaining such axioms we make the following:

3.16. Definition. (1) *A space which is at one and the same time a T_1 and a T_3 space is called a* **regular** *space.*

(2) *A space which is at one and the same time a T_1 and a T_4 space is called a* **normal** *space.*

Before we go one step further, an apology is due for the use of the word "normal" to describe any property of topological spaces, since there is probably no more overworked word in the mathematical lexicon. The word "normal," however, has been hallowed by traditional use for spaces which are both T_1 and T_4, and rather than introduce new terminology, along with new confusion, let's stick to the conventional term.

One more note on terminology: the notion of T_i spaces is due to Alexandroff and Hopf and differs somewhat from what we have done here. The difference is slight and amounts to the following: what we call a regular space, they call a T_3 space, what we call a normal space they call a T_4 space. Some authors follow Alexandroff and Hopf, some do as we have done here; consequently, the reader should be on the lookout, when reading other books, for alternative meanings of the terms "T_3 space" and "T_4 space."

3.17. Theorem. *Each of the following properties of topological spaces is stronger than the next: Normality, Regularity, T_2 (Hausdorff), T_1, T_0, in the sense that if a space satisfies the axioms for any one of these properties, it also satisfies the axioms for all of the following ones as well.*

Proof. Left as an exercise. ∎

We now give some alternative characterizations for regular and normal spaces.

3.18. Theorem. *A T_1 space X is regular if and only if for each $x \in X$ and each $U \in \mathfrak{U}_x$ there is a $V \in \mathfrak{U}_x$, such that $\overline{V} \subseteq U$.*

Proof. Let X be regular, whence axiom T_3 is satisfied. Let $x \in X$, $U \in \mathfrak{U}_x$, then by 1.4 there exists $O \in \mathfrak{O}$, such that $x \in O \subseteq U$. O^c is closed and $x \notin O^c$, thus there exist $O_1, O_2 \in \mathfrak{O}$ such that $x \in O_1$, $O^c \subseteq O_2$, and $O_1 \cap O_2 = \varnothing$. Let $V = O_2^c$, then V is closed, and $x \in O_1 \subseteq O_2^c = V$, since $O_1 \cap O_2 = \varnothing$. Thus since $O_1 \in \mathfrak{O}$, $V \in \mathfrak{U}_x$ by 1.4. Finally since $O^c \subseteq O_2$, $V = O_2^c \subseteq O \subseteq U$, and since V is closed $\overline{V} = V \subseteq U$.

Conversely, let X satisfy the given condition, then since X is T_1 we need only prove that X is T_3. Let $x \in X$, $C \subseteq X$, C closed, $x \notin C$. Then $x \in O = C^c$, O is open, hence $O \in \mathfrak{U}_x$. Select $V \in \mathfrak{U}_x$ and by 1.4 we may assume V open, so that $x \in V \subseteq \overline{V} \subseteq O$. Let $U = \overline{V}^c$, then U is open, $U \cap V = \varnothing$, and $C = O^c \subseteq \overline{V}^c = U$. Thus there exist U, V open, $x \in V$, $C \subseteq U$, and $U \cap V = \varnothing$, whence X is T_3. ∎

3.19. Theorem. *A T_1 space X is normal if and only if for each closed set C and each open set U such that $C \subseteq U$, there exists an open set V such that $C \subseteq V \subseteq \overline{V} \subseteq U$.*

Proof. Left as an exercise in imitating proofs (viz. 3.18). ∎

EXAMPLES (AND EXERCISES)

3.14. Let J be the unit square in the plane with the topology described in Ex. 1.74, then J is a Hausdorff space but not regular, since for the point $x = (\frac{1}{2}, 0)$ and the open set $U = (S_{1/2}(x) \cap \overset{\circ}{J}) \cup \{x\}$, there does not exist a neighborhood V of x such that $\overline{V} \subseteq U$.

3.15. The space given in Ex. 3.14 above is first countable, but not second countable.

3.16. Let $N = \{(x,y) \mid x, y \text{ real}, y \geq 0\}$. Define $\mathfrak{I}$ as follows: If $(x,y) \in \overset{\circ}{N} = \{(x,y) \mid y > 0\}$, define

$$\mathfrak{U}_{(x,y)} = \{U \mid U \supseteq S_\epsilon(x,y),\ S_\epsilon(x,y) \text{ an } \epsilon\text{-sphere about } (x,y),\ y > \epsilon > 0\};$$

if $(x,y) \in \text{Fr}(N) = \{(x,y) \mid y = 0\}$ define

$$\mathfrak{U}_{(x,y)} = \{U \mid U \supseteq [S_\epsilon(x, \epsilon) \cup \{(x,y)\}]\}.$$

Then $(N, 3)$ is a regular space but is not normal.

3.17. Show that a subspace of a T_i space is a T_i space if $i = 1, 2, 3$.

3.18. Prove Theorem 3.17.

3.19. Prove Theorem 3.19.

3.20. Prove that the homeomorphic image of a T_i space is a T_i space if $i = 1, 2, 3, 4$.

The properties of being Hausdorff, normal, and compact are interrelated as is shown in the following:

3.20. Theorem. *A compact Hausdorff space, X, is normal (hence also regular).*

Proof. Since a Hausdorff (i.e., T_2) space is a T_1 space, we need only prove that axiom T_4 is satisfied. We do this in two steps; first we prove axiom T_3 is satisfied. To this end let $x \in X$, $C \subseteq X$, $x \notin C$, and C be closed. C, considered as a subspace of X, is compact by 3.10. For each $y \in C$, there exists a pair of neighborhoods, which we may assume to be open, O_y, O_{xy}, such that $O_y \cap O_{xy} = \emptyset$, $x \in O_{xy}$, $y \in O_y$, since X is Hausdorff. Now $\bigcup_{y \in C} O_y \supseteq C$; thus since C is compact, there exists a finite set $\{y_i \mid i = 1, 2, \ldots, m\}$ such that $C \subseteq \bigcup_{i=1}^{m} O_{y_i}$. Let O_{xy_i} be the open set associated with O_{y_i}, $i = 1, 2, \ldots, m$, and define

$$O = \bigcap_{i=1}^{m} O_{xy_i}, \qquad O' = \bigcup_{i=1}^{m} O_{y_i},$$

then O and O' are open, $C \subseteq O'$, $x \in O$. Also $O \cap O' = \emptyset$, for if

$$z \in O' = \bigcup_{i=1}^{m} O_{y_i},$$

then $z \in O_{y_i}$ for some i, and since $O_{y_i} \cap O_{xy_i} = \emptyset$,

$$z \notin O_{xy_i} \supseteq O,$$

whence $z \notin O$ and $O \cap O' = \emptyset$. Thus axiom T_3 is satisfied.

We now prove axiom T_4 is satisfied. Let $C, D \subseteq X$; C, D closed and disjoint. Since X is T_3, there exists for each $x \in C$ a pair U_x, U'_x of open sets such that $x \in U_x$, $D \subseteq U'_x$, and $U_x \cap U'_x = \emptyset$. Since C is a closed subset of a compact Hausdorff space it is compact, and since $C \subseteq \bigcup_{x \in C} U_x$, there exists a finite set $\{x_i \mid i = 1, 2, \ldots, n\}$ such that $C \subseteq \bigcup_{i=1}^{n} U_{x_i}$. Let

$$U = \bigcup_{i=1}^{n} U_{x_i}, \quad V = \bigcap_{i=1}^{n} U'_{x_i},$$

where U'_{x_i} is the set associated with U_{x_i}. Then as in the first part of the proof, U and V are open and disjoint and $C \subseteq U$, $D \subseteq V$, whence axiom T_4 is satisfied. ∎

The theorem which follows is of a rather special character; it combines conditions on a space and on a mapping sufficient to guarantee a homeomorphism. It is a theorem of occasional utility.

3.21. Theorem. *Let X be a compact space, Y a Hausdorff space, and let $f : X \to Y$ be a one-to-one mapping, then f is a homeomorphism.*

Proof. By Ex. 2.5 we need only prove f is a closed mapping. Let $C \subseteq X$, C closed, then by 3.10 C is compact, whence by 3.7 $f(C)$ is compact and by 3.9 $f(C)$ is closed, since Y is a Hausdorff space, consequently f is a closed mapping. ∎

We have observed (Cf. Ex. 3.17) that every subspace of a T_i space is again a T_i space if $i = 1, 2$, or 3. Thus a subspace of a regular space will also be regular. Unfortunately, however, this will not hold for normal spaces; that is to say, there do exist normal spaces with subspaces that are not normal. It is the object of the following example to demonstrate this. The example is lengthy and somewhat complicated, and is hence worked out in some detail, although a few of the details are left for the reader. The object examined in the example is usually called the Tychonoff plank, since it is "longer" than it is "wide."

EXAMPLES (AND EXERCISES)

3.21. (*Tychonoff plank*) Let S be any uncountable set, and let "$<$" be a well ordering for S, i.e., if $T \subseteq S$, $T \neq \emptyset$, then there is a $t \in T$ such that for all $s \in T$, $t < s$ or $t = s$. The existence of such a well ordering is guaranteed by the axiom of choice, though we do not prove this here. Let Z^+ be the positive integers with the usual ordering and let $Y = S \times Z^+$ be ordered lexicographically by "$<$" [i.e., $(s, n) < (t, m)$ if either $s < t$, or if $s = t$ and $n <' m$ in the usual ordering, $<'$, of the integers]. We observe that Y is well ordered by "$<$." Let $\Omega \in Y$ be the first element of Y which is preceded by uncountably many elements under the lexicographical ordering, let s_2 be the second element of S, and let $\omega = (s_2, 1)$, whence ω is the first element of Y preceded by an infinite number of elements. Finally let

$$X_0 = \{y \mid y \in Y, y \leqq \Omega\}, \quad X_1 = \{y \mid y \in Y, y \leqq \omega\},$$

and provide X_0 and X_1 with the interval topology, i.e., if $x \in X_i$, $i = 0, 1$, let

$$I_{z_1 z_2} = \{y \mid z_1 < y < z_2\}$$

and let

$$\mathcal{U}_x = \{U \mid U \supseteq I_{z_1 z_2} \text{ for some } z_1 < x < z_2\}$$

with suitable modification if x happens to be either the first or last point of X_i, e.g., if x is the last point of X_i, a neighborhood of x is a set which contains an interval of the type $\{y \mid z_i < y \leqq x\}$.

Now we may prove that both X_0 and X_1 are compact and Hausdorff. We leave it to the reader to verify that X_0 and X_1 are Hausdorff and that X_1 is a closed subspace of X_0, and content ourselves here with showing that X_0 is compact. Let $\{O_\alpha\}$ be an open covering of X_0, then there exists $O_0 \in \{O_\alpha\}$ such that $\Omega \in O_0$, and there exists $x_0 \in O_0$ such that the interval $I_{x_0\Omega} \subseteq O_0$. Now define "$x$ is an accessible point" if there exists a finite subcollection of $\{O_\alpha\}$ which contains in its union $I_{x_1 x}$ where x_1 is the first element of X_0, and let

$$A = \{x \mid x_1 \leqq x \leqq x_0, x \text{ is not accessible}\}.$$

Assume $A \neq \emptyset$, then since X_0 is well ordered, A has a first element a, and $a \in O_a \in \{O_\alpha\}$ for suitable choice of O_a. Further there exists $I_{bc} \subseteq O_a$, where $b < a < c$, and since a is the least element of A, $b \notin A$, and b is accessible. There thus exists a finite subfamily of $\{O_\alpha\}$ which contains $I_{x_1 b}$ in its union, consequently this finite subfamily along with O_a contains $I_{x_1 a}$ in its union, and this implies that a is accessible and hence $a \notin A$. This contradiction shows that $A = \emptyset$, whence every x, with $x_1 \leq x \leq x_0$, is accessible; in particular x_0 is accessible. Thus the finite subfamily which contains $I_{x_1 x_0}$ in its union along with the set O_0 covers X_0, and X_0 is compact.

We now consider the space $Z = X_0 \times \dot{X}_1$. Since X_0 and $\dot{X}_1$ are compact, so also is Z by 3.13, and since each of the factor spaces is Hausdorff so also is Z. Thus by 3.20 Z, is normal. It is the space Z that is called the Tychonoff plank. Consider the subspace $Z' = Z - \{(\Omega, \omega)\}$, then Z' is not normal, for if

$$Z_1 = \{z \mid z \in Z, z = (x, \omega), \text{ where } x \text{ is arbitrary}\}$$

(i.e., the top edge of the plank) and

$$Z_2 = \{z \mid z \in Z, z = (\Omega, x), \text{ where } x \text{ is arbitrary}\}$$

(i.e., the right-hand edge of the plank), then Z_1 and Z_2 are closed subsets of Z, being the inverse image of points under projection mappings. Thus $Z_1' = Z_1 \cap Z'$ and $Z_2' = Z_2 \cap Z'$ are closed subsets of Z' and are furthermore disjoint, since $Z_1 \cap Z_2 = (\Omega, \omega) \notin Z'$. Now let O_1' and O_2' be arbitrary open sets of Z' such that $O_1' \supseteq Z_1'$ and $O_2' \supseteq Z_2'$. Consider O_2', the set which covers the right-hand end of the plank; O_2' is the union of basic sets, i.e., rectangles of the form $I_{x_\alpha \Omega} \times I_{x_1 x_2}$, where $x_\alpha < \Omega$, $x_1 < x_2$. Now the set of x_α where x_α is the initial point of the interval $I_{x_\alpha \Omega}$ is countable, hence has an upper bound $x < \Omega$, by Ex. 3.22 below. Consider the point $(x, \omega) \in Z_1'$, then $(x, \omega) \in O_1'$ and there exists a rectangle

$$R = I_{x_3 x_4} \times I_{x_5 \omega} \qquad \text{where } x_3 < x < x_4, \qquad x_5 < \omega,$$

such that $(x, \omega) \in R \subseteq O_1'$. It follows (Cf. Ex. 3.23 below) that

$$R \cap O_2' \neq \emptyset, \qquad \text{whence } O_1' \cap O_2' \neq \emptyset.$$

From this it follows that Z' is not normal.

3.22. Prove that any countable set of $x_\alpha \in X_0$, with $x_\alpha < \Omega$ for each α, has a least upper bound $x < \Omega$.

SEC. 3.2   SPECIAL TYPES OF TOPOLOGICAL SPACES   87

3.23. Prove that $R \cap O_2' \neq \emptyset$ in Ex. 3.21 above.

3.24. In the proof of compactness of X_0 in Ex. 3.21 above, why need we only consider points $a \in A$ such that $x_1 < a < \Omega$?

NOTE: The preceding description of the Tychonoff plank is really much more complicated than it need be. The reason is that we have tried to avoid introducing the notion of ordinal numbers. To the student familiar with ordinal numbers it should be clear that the space X_0 above is simply the set of all ordinal numbers not greater than the first uncountable ordinal, Ω, while X_1 is the set of all ordinal numbers not greater than the first infinite ordinal, ω.

The Tychonoff plank motivates the definition of a sort of space whose every subspace is normal, namely

3.22. Definition. *A space is said to be **completely normal** if and only if every one of its subspaces is normal.*

We would now like to relate completely normal spaces to some other properties, but we first require a result, which is usually called the Lindelöf theorem:

3.23. Theorem. *Let X be a space with a countable basis, let*

$$\{O_\alpha \mid \alpha \in A\}$$

be an open covering of $Y \subseteq X$, then there exists a countable subcovering $\{O_i'\} \subseteq \{O_\alpha\}$.

Proof. With each $x \in Y$ and each O_α associate $B_{\alpha,x} \in \mathcal{B}$ such that $x \in B_{\alpha,x} \subseteq O_\alpha$. Define $\{B_\alpha\} = \{B_{\alpha,x} \in \mathcal{B} \mid x \in Y, \alpha \in A\}$, then since $\mathcal{B}$ is countable and $\{B_\alpha\} \subseteq \mathcal{B}$, $\{B_\alpha\}$ is likewise countable, i.e., $\{B_\alpha\} = \{B_i \mid i = 1, 2, \ldots\}$, and we reindex the B_α in this way using the positive integers. With each B_i associate O_i' such that $B_i \subseteq O_i' \in \{O_\alpha\}$, then $\{O_i' \mid i = 1, 2, \ldots\}$ is the desired countable covering, for clearly $\{O_i'\}$ is countable, $\{O_i'\} \subseteq \{O_\alpha\}$, and if $x \in Y$, $x \in O_\alpha$ for some α, whence $x \in B_{\alpha,x} = B_\alpha = B_i$ for some i. Thus $x \in B_i \subseteq O_i'$ and $\bigcup_{i=1}^{\infty} O_i' \supseteq Y$. ∎

Having proved the Lindelöf theorem we are now in a position to establish the following:

3.24. Theorem. *Every regular space with a countable basis is completely normal.*

Proof. We remark that by Ex. 1.72 and by Ex. 3.17 any subspace of a regular space with a countable basis is again such a space. Thus all we need to prove is that any regular space with a countable basis is normal.

To this end let C and D be closed disjoint subsets of X. For each $x \in C$, there exists an open set U_x such that $x \in U_x \subseteq \overline{U_x} \subseteq D^c$, i.e., $\overline{U_x} \cap D = \varnothing$. Then $\bigcup_{x \in C} U_x \supseteq C$ and by 3.23 we may select a countable subset $\{U_i \mid i = 1, 2, \ldots\} \subseteq \{U_x \mid x \in C\}$ such that

$$\bigcup_{i=1}^{\infty} U_i \supseteq C.$$

Similarly select $\{V_i \mid i = 1, 2, \ldots\}$ such that

$$\overline{V_i} \cap C = \varnothing, \quad \bigcup_{i=1}^{\infty} V_i \supseteq D.$$

Now define $O_1 = U_1$, $W_1 = V_1$, and inductively thereafter define

$$O_n = U_n \cap \left(\bigcap_{i=1}^{n} \overline{V_i}^c\right) \quad \text{and} \quad W_n = V_n \cap \left(\bigcap_{i=1}^{n} \overline{U_i}^c\right).$$

Clearly O_n and W_n are open, whence

$$O = \bigcup_{n=1}^{\infty} O_n \quad \text{and} \quad W = \bigcup_{n=1}^{\infty} W_n$$

are open. Further since $C \subseteq \overline{V_i}^c$ for each i, and $C \subseteq \bigcup_{i=1}^{\infty} U_i$ we have $C \subseteq O$ and similarly $D \subseteq W$. Also we observe that if $n \geq m$,

$$O_n \cap W_m = U_n \cap \left(\bigcap_{i=1}^{n} \overline{V_i}^c\right) \cap V_m \cap \left(\bigcap_{i=1}^{m} \overline{U_i}^c\right) \subseteq \overline{V_m}^c \cap V_m = \varnothing,$$

and if $n \leq m$,

$$O_n \cap W_m \subseteq U_n \cap \overline{U_n}^c = \varnothing.$$

Thus for each m, $O_n \cap W_m = \varnothing$ whence $O_n \cap W = \varnothing$, and since this last holds for each n, we have further that $O \cap W = \varnothing$. We have thus constructed the required disjoint open sets which contain D and C, and X is normal. ∎

EXAMPLES (AND EXERCISES)

3.25. Show that every completely normal space is normal.

3.26. Show that every regular space with a countable basis is normal.

3.27. Show that every subspace of a completely normal space is again completely normal.

§3 Countable Compactness

We observed some time ago that the notion of compactness was motivated by the Heine-Borel theorem for the real line. Another property of the reals, the so-called Bolzano-Weierstrass property, asserts, in one version, that any closed bounded infinite set of reals has a limit point which belongs to the set. This property is the motivating one for countable compactness, according to the following:

3.25. Definition. *Let X be a topological space, then X is **countably compact** if and only if every infinite subset of X has a limit point in X.*

We immediately remark that the notion of countable compactness is no stronger than the notion of compactness. This is shown by the following:

3.26. Theorem. *Let X be a compact space, then X is countably compact.*

Proof. Let $A \subseteq X$, A infinite, and suppose that A has no limit point in X. Select from A an infinite sequence $\{x_n\}$ of distinct points, then by assumption $\{x_n\}$ has no limit point in X. Thus for each n, there exists an open $V_n \in \mathfrak{U}_{x_n}$ such that

$$V_n \cap \{x_n\} = x_n.$$

Further, for each $y \notin \{x_n\}$, since y is not a limit point of $\{x_n\}$, there exists an open $V_y \in \mathfrak{U}_y$ such that $V_y \cap \{x_n\} = \emptyset$. Let

$$V_0 = \bigcup_{y \notin \{x_n\}} V_y,$$

then V_0 is open. Now

$$\bigcup_{i=0}^{\infty} V_i = X,$$

yet clearly no finite subcollection of this open covering covers X, since the omission of any V_i, $i \geq 1$ implies the omission of x_i. This, then, implies that X is not compact, and this contradiction proves that X is countably compact. ∎

It is in fact the case that countable compactness is strictly weaker than compactness, in the sense that there exist countably compact spaces which are not compact. This is shown in the following examples.

EXAMPLES (AND EXERCISES)

3.28. Let X be the set of positive integers. Let $B_i = \{2i, 2i - 1\}$ and let $\mathfrak{B} = \{B_i \mid i = 1, 2, \ldots\}$ be a basis for a topology for X, then X is countably compact but not compact.

3.29. If X is a space such that every countable open cover of X contains a finite subcover of X, then X is countably compact. (This result motivates the name "countable compactness.")

3.30. Show that although the space X of Ex. 3.28 above is countably compact it does not have the property that every countable open cover has a finite subcover.

3.31. Show that any closed subspace of a countably compact space is again countably compact.

There is an unfortunate state of affairs exhibited in the preceding exercises (Ex. 3.29, 3.30), namely that though every space which has the property that every countable open cover has a finite subcover is countably compact, the converse of this statement does not hold. It may consequently seem strange that we motivate the nomenclature of countably compact spaces as we do (Cf. note after Ex. 3.29 above). Our reason for doing so is that the simple addition of the hypothesis that X be T_1 makes the two properties equivalent, as is shown by the following:

3.27. Theorem. *Let X be a T_1 space, then X is countably compact if and only if every countable open cover of X contains a finite subcover.*

Proof. Exercise 3.29 above proves this in one direction. We must thus only prove that countable compactness implies the given condition. To this end, let $\{O_n\}$ be a countable family of open sets which cover X. Suppose no finite subcollection covers X, then if $C_n = X - \bigcup_{i=1}^{n} O_i$, $C_n \neq \emptyset$ for any n. Furthermore C_n, as the complement of an open set, is closed, and also $C_n \supseteq C_{n+1}$. We can thus, for each n, select a point $x_n \in C_n$. If the set $S = \{x_n \mid n = 1, 2, \ldots\}$ is infinite, then since X is countably compact, S has a limit point x_0 in X. By Ex. 3.12, x_0 is a limit point of

$$S_k = \{x_n \mid n = k, k+1, \ldots\} \subseteq C_k.$$

Since C_k is closed, $x_0 \in C_k$ for each k, whence $x_0 \notin O_k$ for any k. This contradicts the fact that $\{O_n\}$ covers X. It thus follows that the set $S = \{x_n \mid n = 1, 2, \ldots\}$ is finite. Thus there exist x' and $N > 0$ such that for all $n \geq N$, $x_n = x'$, and as before $x' \in C_n$ for $n > N$, and because the C_n are a decreasing sequence of sets (i.e., $C_n \supseteq C_{n+1}$), $x' \in C_n$ for all n. Once again this means that $x' \notin O_n$ for all n, and $\{O_n\}$ is not a covering. This contradiction establishes the theorem. ∎

We show now that under suitable conditions it is in fact the case that countable compactness implies compactness.

3.28. Theorem. *A second countable T_1 space is countably compact if and only if it is compact.*

Proof. Compactness implies countable compactness by 3.26. Conversely if X is countably compact, and $\{O_\alpha\}$ is an open covering of X, then by 3.23 there is a countable subfamily $\{O_i\}$ of $\{O_\alpha\}$ that also covers X. Then by 3.27 $\{O_i\}$ contains a finite subcover, whence X is compact. ∎

EXAMPLES (AND EXERCISES)

3.32. Show that a T_1 space is countably compact if and only if every countable family of closed sets that has the finite intersection property has a nonvacuous intersection.

3.33. A T_1 space is countably compact if and only if every infinite open cover (where infinite refers to the indexing set) has a proper subcover. [*Hint*. Imitate 3.27.]

§ 4. Local Compactness

3.29. Definition. *A space X is said to be **locally compact** provided that for each $x \in X$ there exists $U \in \mathcal{U}_x$ such that U is compact.*

EXAMPLES (AND EXERCISES)

3.34. The real line with the usual topology is locally compact but not compact.

3.35. The real plane with the usual topology is locally compact but not compact.

3.36. Every compact space is locally compact.

3.37. Every closed subspace of a locally compact space is again locally compact.

The motivation for introducing the notion of local compactness is twofold. First of all, as has already been pointed out (Cf. Ex. 3.34, 3.35), the real line and the real plane are locally compact, as is, in fact,

$$E^n = \underset{i=1}{\overset{n}{\times}} R_i,$$

where each R_i is just the reals with the usual topology. Since such spaces are of general interest to the analyst, the notion of local compactness is of some concern in analysis. Second of all, there is a very pretty topological analog of the process of stereographic projection. To refresh the memories of those of us who may have forgotten, stereographic projection consists of the following:

SEC. 3.4 SPECIAL TYPES OF TOPOLOGICAL SPACES

Consider the real plane with coordinate x and y axes. Construct a sphere of radius $1/2$ tangent to the plane at $(0, 0)$, and let the real three-dimensional space with coordinate axes be so located that the origin is at the origin of the real plane, the x' and y' axes of the real three-dimensional space coincide with the x and y axes of the real plane respectively, and the positive z'-axis passes through the north pole of the sphere, the south pole being at $(0, 0)$. Now to each point $P = (x, y)$ of the real plane we can make correspond a point $Q = (x', y', z')$ of the sphere S, whose equation is given by $x'^2 + y'^2 + z'^2 = z'$, in the following fashion: Draw a straight line from $N = (0, 0, 1)$, the north pole, to $P = (x, y, 0)$. This line meets the sphere in some point; let this point be Q. A simple calculation shows that

$$x' = \frac{x}{1 + r^2}, \quad y' = \frac{y}{1 + r^2}, \quad z' = \frac{r^2}{1 + r^2},$$

where $r^2 = x^2 + y^2$.

It is not difficult to see that the mapping $f : R \to S - \{(0, 0, 1)\}$ that carries the real plane into the sphere whose north pole has been removed, which is defined by

$$f((x, y)) = \left(\frac{x}{1 + r^2}, \frac{y}{1 + r^2}, \frac{r^2}{1 + r^2}\right),$$

where $r^2 = x^2 + y^2$, is one-to-one and onto, for another calculation shows that

$$f^{-1}((x', y', z')) = \left(\frac{x'}{1 - z'}, \frac{y'}{1 - z'}\right),$$

and further from this it is seen that both f and f^{-1} are continuous. Thus f is a homeomorphism.

Now what happens is this: We have embedded the locally compact real plane in the compact sphere S in a homeomorphic fashion so that $S - f(R)$ is a single point $N = (0, 0, 1)$. In some sense, then, it is rather natural to ask, "If X is a locally compact space, which is not compact, is it possible to find a compact space X' and a homeomorphism $f : X \to X'$ such that $X' - f(X)$ is a single point?" The answer to this question is, in fact, "Yes," and we shall set out in the next few theorems to prove exactly this result. All that this shows is that stereographic projection is really only one instance of a rather general phenomenon.

It might be pointed out here that the utility of this result lies in the fact that we can view a locally compact space as a subspace of a compact space X' and, in general, compact spaces are more pleasant to deal with than noncompact spaces.

3.30. Definition. *Let X be a noncompact T_1 space, let $\mathcal{K}$ be the family of closed compact subsets of X, and let ι be an ideal element such that $\iota \notin X$. Define a space $^cX = X \cup \{\iota\}$ with basis*

$$\mathcal{B} = \{U \mid U \in \mathcal{O}, \text{ or } U = {}^cX - K, K \in \mathcal{K}\},$$

*then ^cX is called the **one-point compactification** of X.*

The insistence that X be a T_1 space is motivated by the desire to have $\mathcal{K} \neq \emptyset$. For if $\mathcal{K} = \emptyset$, ι would have no neighborhood system. The most important properties of the one-point compactification, ^cX, are exhibited in the following theorems.

3.31. Theorem. *Let X be a noncompact T_1 space, then ^cX is a T_1 space.*

Proof. Clearly $\bigcup_{B \in \mathcal{B}} B = {}^cX$. Let $x \in {}^cX$, and let $U, V \in \mathcal{B}$ such that $x \in V$ and $x \in U$. If $x \in X$, U and V are open and consequently there exists $W = U \cap V$ open in X, hence again in $\mathcal{B}$ such that

$$x \in W \subseteq U \cap V.$$

If $x = \iota$, then

$$U = {}^cX - K_1, \quad V = {}^cX - K_2, \quad K_1, K_2 \in \mathcal{K}.$$

Let

$$W = U \cap V = ({}^cX - K_1) \cap ({}^cX - K_2) = {}^cX - (K_1 \cup K_2),$$

and $K_1 \cup K_2$ is again closed and compact, whence $W \in \mathcal{B}$, and $x \in W \subseteq U \cap V$. Consequently, by 1.29, ^cX is a topological space.

We show now that ^cX is T_1. Let $x, y \in {}^cX$, $x \neq y$. If $x, y \in X$ then since X is T_1, there exists $U \in \mathcal{U}_x$ (or $\mathcal{U}_y$) such that $y \notin U$ (or $x \notin U$). On the other hand, if one of x and y (say x) is ι, then since $\{y\}$ is a closed compact set in X, $U = {}^cX - \{y\} \in \mathcal{U}_x$ and $y \notin U$. In either case ^cX is a T_1 space. ∎

3.32. Theorem. *Let X be a noncompact T_1 space, then ^cX is compact.*

Proof. Let $\{O_\alpha\}$ be an open covering of ^cX, then some one of the covering sets contains ι, let us call it O_0. There exists then a basic set $B \subseteq O_0$, and $B = {}^cX - K$, K closed and compact in X. Now

$$K \subseteq \bigcup_{\alpha \in A} O_\alpha,$$

thus there exists a finite subcollection $\{O_i \mid i = 1, 2, \ldots, n\}$ such that

$$\bigcup_{i=1}^n O_i \supseteq K.$$

Since $^cX - K \subseteq O_0$, we have

$$^cX \subseteq \bigcup_{i=0}^n O_i,$$

and we have found a finite subcovering of the covering $\{O_\alpha\}$, whence ^cX is compact. ∎

3.33. Theorem. *Let X be a noncompact T_1 space, ^cX its one-point compactification, then X is locally compact and Hausdorff if and only if ^cX is Hausdorff.*

Proof. Let ^cX be Hausdorff, then X as a subspace of ^cX (Cf. Ex. 3.39 below) is again Hausdorff. Let $x \in X$, then there exist open sets U and V of ^cX such that $x \in U$, $\iota \in V$, $U \cap V = \emptyset$, $V \supseteq {}^cX - K$ for some $K \in \mathcal{K}$. Now

$$({}^cX - K) \cap U \subseteq V \cap U = \emptyset$$

so that $U \subseteq K$. Since K is closed and compact, $\bar U \subseteq K$, and $\bar U$ is compact. Clearly $\bar U \in \mathfrak{U}_x$, whence X is locally compact.

Conversely let X be locally compact and Hausdorff. Let $x, y \in {}^cX$, $x \neq y$. If $x, y \in X$, there exist U, V open in X, hence also open in ^cX such that $x \in U$, $y \in V$, $U \cap V = \emptyset$. If one of x or $y = \iota$, say $x = \iota$, then since X is locally compact we may select $U \in \mathfrak{U}_y$ such that U is compact, hence also closed. Then

$$^cX - U \in \mathfrak{U}_\iota \quad \text{and} \quad ({}^cX - U) \cap U = \emptyset.$$

Thus in any case ^cX is a Hausdorff space. ∎

EXAMPLES (AND EXERCISES)

3.38. (a) Show that the mapping $f: X \to {}^cX$ defined by $f(x) = x$ is a homeomorphism of X with $f(X)$.

(b) Infer from (a) above that every locally compact Hausdorff space is homeomorphic to a subspace of a compact Hausdorff space.

3.39. It is clear from 3.30 that X is a subset of ${}^cX$. Show further that it is also a subspace.

3.40. Show that every locally compact Hausdorff space is regular. (*Hint.* Use 3.33, 3.20, Ex. 3.17, and Ex. 3.39 above.)

3.41. Let $X = \{n \mid n > 0, n \text{ an integer}\}$, define

$$S_{n,\epsilon} = \{m \mid |1/n - 1/m| < \epsilon\}$$

and let $\mathfrak{B} = \{S_{n,\epsilon} \mid n = 1, 2, \ldots, \epsilon > 0\}$. Show that with $\mathfrak{B}$ as basis, X is a topological space, X is Hausdorff, and that ${}^cX$ is obtained by adjoining the point $\iota = \infty$ and defining the basis elements at ∞ by $S_{\infty,\epsilon} = \{m \mid |1/m| < \epsilon\}$.

3.42. (*Term paper*) Paracompactness.

Definition. *A family $\{F_\alpha\}$ of subsets of a topological space X is **locally finite** if and only if for each $x \in X$ there exists $U \in \mathfrak{U}_x$ such that $U \cap F_\alpha \neq \varnothing$ for only finitely many indices α.*

Definition. *Let $\{U_\alpha\}$ be a covering of a topological space, then a covering $\{V_\beta\}$ is said to be a **refinement** of $\{U_\alpha\}$ if for each β there exists an α such that $V_\beta \subseteq U_\alpha$.*

Definition. *A topological space is said to be **paracompact** if it is a Hausdorff space, and if each open cover $\{O_\alpha\}$ has a locally finite refinement.*

Prove at least the following:

(1) Every compact Hausdorff space is paracompact, but not conversely.
(2) Paracompactness is a topological property.
(3) Every paracompact space is regular.
(4) Every paracompact space is normal.

(5) Every closed subspace of a paracompact space is paracompact.
(6) The topological product of a paracompact space and a compact space is paracompact.
(7) Show by counterexample that the topological product of two paracompact spaces need not be paracompact.

NOTE: A result owing to A. H. Stone assures us that every metric space (Cf. Chapter 5) is paracompact. This result appears in *Bull. Amer. Math. Soc.* (1948), pp. 977–982.

CHAPTER 4

FURTHER SPECIAL TYPES OF TOPOLOGICAL SPACES (MOSTLY VARIETIES OF CONNECTEDNESS)

§1 Introduction

In this chapter we continue to explore various restrictions we can place on the topology of a space. The major notion we shall explore is that of connectedness. Intuitively, a connected space is one which consists of one piece. The problem then is to write a formal definition of the intuitive notion of "one-pieceness." Let us first consider a few examples. Certainly we would think of R, the real line, and E, the real plane, as being of one piece. Suppose we consider $R - \{0\}$; surely this is not of one piece, since without doing any serious violence to the topological (i.e., geometric) structure we can separate $R - \{0\}$ into the two pieces

$$R' = \{x \mid x \in R, x > 0\} \quad \text{and} \quad R'' = \{x \mid x \in R, x < 0\}.$$

Let us consider one further example, namely

$$X = \left\{(x, y) \mid x, y \in R, y = \sin\frac{1}{x} \text{ for } 0 < x \leq \frac{1}{\pi}\right.$$
$$\left. \text{or } x = 0, \text{ and } -1 \leq y \leq 1\right\}.$$

Certainly the infinite arc A defined by

$$y = \sin\frac{1}{x} \quad \text{for} \quad 0 < x \leq \frac{1}{\pi}$$

is all of one piece, and certainly the line segment

$$L = \{(0, y) \mid -1 \leqq y \leqq 1\}$$

is of one piece. Do we do violence to the topological structure of X by separating these two pieces, A and L? It would certainly appear so, since each point of L is a limit point of points of A, e.g., the point $(0, \frac{1}{2})$ is a limit point of the set

$$A_{1/2} = \left\{(x, y) \mid x = \frac{1}{\pi/6 + 2n\pi}, y = \sin\frac{1}{x}, n = 1, 2, \ldots\right\} \subseteq A.$$

It would thus appear that we don't want a connected space to be such that we can write it as the union of two sets, neither of which has a limit point of the other. With this in mind we make the following two definitions.

4.1. Definition. *Let X be a topological space, A and B nonempty subsets of X, then A and B are said to be **separated** if and only if $A \cap \overline{B} = \emptyset$ and $\overline{A} \cap B = \emptyset$.*

4.2. Definition. *Let X be a topological space, then X is said to be **connected** provided that X cannot be written as the union of two nonempty separated sets. A set $A \subseteq X$ is said to be connected if A is a connected subspace of X.*

§ 2 Connected Spaces

There are a number of other ways in which a connected space may be defined. Several of these are given in the following:

4.3. Theorem. *A space X is connected if and only if any one of the following conditions is met:*

(1) *X is not the disjoint union of two nonempty open sets.*
(2) *X is not the disjoint union of two nonempty closed sets.*
(3) *The only sets in X which are both open and closed are $\emptyset$ and X.*
(4) *For every continuous $f: X \to R$, $f(X)$ does not consist of two distinct real numbers.*

Proof. We leave the proofs of (1), (2), and (3) as exercises. We prove (4) contrapositively. Let f be a continuous function of X into R such that $f(X) = \{a\} \cup \{b\}$, $a \neq b$. Let $\epsilon = \frac{1}{3}|a - b|$ and define

$$U = (a - \epsilon, a + \epsilon), \qquad V = (b - \epsilon, b + \epsilon),$$

then $U \cap V = \emptyset$, and U and V are open in R. Thus $f^{-1}(U)$ and $f^{-1}(V)$ are open in X since f is continuous, and

$$X = f^{-1}(U) \cup f^{-1}(V),$$

and thus by part (1), X is not connected.

Conversely, suppose X is not connected, then there exist open sets U, V, such that $X = U \cup V$, $U \cap V = \emptyset$, by part (1). Define $f : X \to R$ by $f(x) = 0$ if $x \in U$, $f(x) = 1$ if $x \in V$, then f is clearly continuous. ∎

Connectedness is a topological property, in fact even more is true, namely:

4.4. Theorem. *Let X be connected, let $f : X \to Y$ be continuous and onto, then Y is connected.*

Proof. We proceed contrapositively, by proving that if Y is not connected, then X is not connected. If Y is not connected, we can by 4.3(1) write $Y = A \cup B$, where $A \cap B = \emptyset$ and A and B are open and nonempty in Y. Then

$$X = f^{-1}(A) \cup f^{-1}(B)$$

where, since f is continuous, $f^{-1}(A)$ and $f^{-1}(B)$ are open in X, and $f^{-1}(A) \cap f^{-1}(B) = \emptyset$. Thus X is not connected. ∎

EXAMPLES (AND EXERCISES)

4.1. Let $C \subseteq R$, the reals, and let C be connected. Let $a, b \in C$, and let $a < c < b$, then $c \in C$. [*Hint.* Suppose $c \notin C$, define $f : C \to R$ by $f(x) = a$ if $x < c$ and $f(x) = b$ if $x > c$, then prove f is continuous and use 4.3(4).]

4.2. The nonempty connected subsets of R are of the following forms:

(a) Intervals or points, i.e., (a, b), $[a, b)$, $(a, b]$, $a < b$, or $[a, b]$, $a \leq b$.

(b) Rays, i.e., (a, ∞), $[a, \infty)$, $(-\infty, a)$, or $(-\infty, a]$.

(c) R itself.

4.3. Let X be a space, then X is connected if and only if for each $f: X \to R$, f continuous, f has the intermediate value property, i.e., if $f(x_1) = a$, $f(x_2) = b$, and c is between a and b, then there exists $x \in X$ such that $f(x) = c$.

4.4. Prove Theorem 4.3(1), (2), (3).

4.5. Theorem. *Let X be a space, $X = X_1 \cup X_2$, where X_1 and X_2 are separated. Let $C \subseteq X$ be connected, then either $C \subseteq X_1$ or $C \subseteq X_2$.*

Proof. Define $C_1 = C \cap X_1$, $C_2 = C \cap X_2$, then $C = C_1 \cup C_2$. Then $\overline{C_1} \subseteq \overline{X_1}$ and $\overline{C_2} \subseteq \overline{X_2}$, whence

$$C_1 \cap \overline{C_2} \subseteq X_1 \cap \overline{X_2} = \varnothing \quad \text{and} \quad \overline{C_1} \cap C_2 \subseteq \overline{X_1} \cap X_2 = \varnothing.$$

Thus if both $C_1 \neq \varnothing$ and $C_2 \neq \varnothing$, C is not connected, consequently either $C_1 = \varnothing$ in which case $C \subseteq X_2$ or $C_2 = \varnothing$ in which case $C \subseteq X_1$. ∎

4.6. Definition. *Let $\mathfrak{D} = \{D_\alpha\}$ be a family of sets in some set X. A finite subset*

$$\mathfrak{D}_n = \{D_i \mid i = 1, 2, \ldots, n\} \subseteq \mathfrak{D}$$

*is a **chain** provided that $D_i \cap D_{i+1} \neq \varnothing$, $i = 1, 2, \ldots, n-1$. Such a chain is said to be **simple** if $D_i \cap D_j = \varnothing$ unless $j = i \pm 1$. The family of subsets $\mathfrak{a} = \{A_\beta\}$ of X is said to be **chained by** $\mathfrak{D}$ if for each $A_\beta, A_\gamma \in \mathfrak{a}$, there is a chain $\mathfrak{D}_n \subseteq \mathfrak{D}$, such that $A_\beta \cap D_1 \neq \varnothing$, $A_\gamma \cap D_n \neq \varnothing$. $\mathfrak{a}$ is said to be **simply chained by** $\mathfrak{D}$ if for each $A_\beta, A_\gamma \in \mathfrak{a}$ there is a simple chain $\mathfrak{D}_n \subseteq \mathfrak{D}$ such that $A_\beta \cap D_1 \neq \varnothing$, $A_\gamma \cap D_n \neq \varnothing$.*

4.7. Theorem. *Let X be a connected topological space, $\{O_\alpha\}$ an open covering of X, then $\{O_\alpha\}$ is chained by $\{O_\alpha\}$.*

Proof. We proceed contrapositively. Let $\{O_\alpha\}$ be an open covering of X which is not chained by $\{O_\alpha\}$. Let $O_1 \in \{O_\alpha\}$ and let

$$\mathfrak{O}_1 = \{O \mid O \in \{O_\alpha\}, \text{ such that } \{O, O_1\} \text{ is chained by } \{O_\alpha\}\}.$$

$\mathcal{O}_1 \neq \emptyset$ since $O_1 \in \mathcal{O}_1$. Let $\mathcal{O}_2 = \{O_\alpha\} - \mathcal{O}_1$, then $\mathcal{O}_2$ is not empty, since $\{O_\alpha\}$ is not chained by $\{O_\alpha\}$. Let
$$U = \bigcup_{O \in \mathcal{O}_1} O, \quad V = \bigcup_{O \in \mathcal{O}_2} O,$$
then $U \cap V = \emptyset$, U and V are each nonempty, whence $X = U \cup V$ is not connected. ∎

4.8. Theorem. *Let X be a topological space, then X is connected if and only if X is simply chained by any open covering of X, i.e., any two points of X can be joined by a finite set of open sets of the given covering.*

Proof. Let X be connected, $\{O_\alpha\}$ an open covering of X, and let $x \in X$. Let $C_x = \{y \mid \{x, y\}$ is simply chained by $\{O_\alpha\}\}$. $C_x \neq \emptyset$, since $x \in C_x$. We show that C_x is both open and closed. Let $y \in C_x$, then there exists a simple chain
$$\mathcal{O}_n = \{O_i \mid i = 1, 2, \ldots, n\}$$
such that $x \in O_1$, $y \in O_n$. Now for any $z \in O_n$, $\mathcal{O}_n$ is a simple chain joining x to z, thus $z \in C_x$, and $O_n \subseteq C_x$, whence C_x is open, since O_n is a neighborhood of y.

Now let y be a limit point of C_x, then $y \in O_\alpha$ for some α and $O_\alpha \cap C_x \neq \emptyset$. Thus let $z \in O_\alpha \cap C_x$, and let
$$\mathcal{O}_n = \{O_i \mid i = 1, 2, \ldots, n\}$$
be a simple chain from x to z. Further let k be the smallest index $1 \leq k \leq n$ such that $O_\alpha \cap O_k \neq \emptyset$. Such a k surely exists since $O_n \cap O_\alpha \neq \emptyset$, then
$$\mathcal{O}_{k+1} = \{O_i \mid i = 1, 2, \ldots, k, \alpha\}$$
is a simple chain from x to y, and $y \in C_x$. Thus C_x is closed. The proof is completed by invoking 4.3(3).

Conversely, suppose X is not connected, then
$$X = A \cup B, \quad A \neq \emptyset, \quad B \neq \emptyset, \quad A \cap B = \emptyset,$$
A and B open. Let $a \in A$, $b \in B$, then $\{A, B\}$ is an open covering and there is no simple chain (in fact no chain of any kind) from a to b

consisting of sets of the open covering $\{A, B\}$. This is a contradiction, hence X is connected. ∎

4.9. Theorem. *Let X be a space, let N be a connected set in X, and let $N \subseteq A \subseteq \overline{N}$, then A is connected, i.e., under the adjunction of any number of limit points a connected set remains connected.*

Proof. Suppose A is not connected, then

$$A = A_1 \cup A_2, \quad A_1 \neq \emptyset, \quad A_2 \neq \emptyset,$$

A_1 and A_2 separated. Since N is connected, and $N \subseteq A$, either $N \subseteq A_1$ or $N \subseteq A_2$ by 4.5. We assume $N \subseteq A_1$. Since N is connected and A is not, $A - N \neq \emptyset$, and further $A_2 \neq \emptyset$, thus we may select $x \in A - N$ such that $x \in A_2$. Then since $A \subseteq \overline{N}$,

$$x \in \overline{N} \subseteq \overline{A}_1,$$

whence $\overline{A}_1 \cap A_2 \neq \emptyset$, and this contradicts the fact that A_1 and A_2 are separated. This contradiction establishes that A is connected. ∎

EXAMPLES (AND EXERCISES)

4.5. A space X is said to be dense in itself if each $x \in X$ is a limit point of $X - \{x\}$. Show that every connected T_1 space which consists of more than one point is dense in itself.

4.6. Show that in Ex. 4.5 above X must be a T_1 space for the conclusion to hold by considering

$$X = \{a, b\} \quad \text{where } \mathcal{O} = \{\{a, b\}, \{b\}, \emptyset\}.$$

4.7. Show that in a connected T_1 space every connected set which consists of more than one point is infinite.

4.8. Prove that if X is a connected space,

$$\mathcal{C} = \{C_i \mid i = 1, 2, \ldots, n\}$$

a finite covering of X by closed sets C_i, then $\mathcal{C}$ is chained by $\mathcal{C}$.

4.9. Let X be a space, $M, N \subseteq X$, M, N connected, then if either

   (a) $M \cap N \neq \emptyset$ or   (b) $M \cap \overline{N} \neq \emptyset$

then $M \cup N$ is connected.

4.10. If X is a space, $N \subseteq M \subseteq X$, M and N connected, and if $M - N = A \cup B$, where A and B are separated, then $N \cup A$ is connected.

4.11. Suppose we had proved the following in place of 4.9:

Theorem 4.9(a). *If N is a connected set in X, then so also is $\overline{N}$.*

Prove 4.9 on the basis of 4.9(a) by considering A with the relative topology.

4.12. Let $\{C_\alpha\}$ be a family of connected sets in some space indexed by $\alpha \in A$. If
$$\bigcap_{\alpha \in A} C_\alpha \neq \emptyset,$$
then $\bigcup_{\alpha \in A} C_\alpha$ is connected.

4.13. Let $\{A_i \mid i = 1, 2, \ldots, n\}$ be a finite family of connected sets such that
$$A_i \cap A_{i+1} \neq \emptyset, \quad i = 1, 2, \ldots, n - 1.$$
Then $\bigcup_{i=1}^{n} A_i$ is connected.

4.14. Let $A \subseteq X$ be a connected set, and let $B \subseteq X$. Suppose $B \cap A \neq \emptyset$, and $B^c \cap A \neq \emptyset$. Show that $A \cap \operatorname{Fr}(B) \neq \emptyset$, where $\operatorname{Fr}(B)$ is the frontier of B (Cf. 1.16). Infer that in a connected space, if $B \neq \emptyset$, $B \neq X$, then $\operatorname{Fr}(B) \neq \emptyset$.

4.15. Let A and B be closed subsets of the space X. If $A \cup B$ and $A \cap B$ are connected, so also are A and B. Show by counterexample that both A and B must be closed, i.e., if one of A and B is not closed then in general the theorem fails.

§ 3 Components

In some topological spaces it is occasionally interesting to examine the maximal connected subsets. We can do this via certain subsets of the space called components, as defined in the following:

4.10. Definition. *Let X be a space, $M \subseteq X$ and $x \in M$, then C_x, the* **component of x in M,** *is the union of all connected subsets of M to which x belongs, or more formally*

$$C_x = \{y \mid x \text{ and } y \text{ lie in a connected subset of } M\}.$$

4.11. Theorem. *Let X be a space, C_x a component of M in X, then*

(1) C_x *is connected.*
(2) C_x *is maximal with respect to the properties:*
   (a) $x \in C_x$,
   (b) C_x *is a connected subset of M, i.e., if $A \subseteq M$, $x \in A$ and A is connected, then $A \subseteq C_x$.*

Proof. (1) Let $y \in C_x$, define A_y to be a connected subset of M such that $x, y \in A_y$. Now $\bigcup_{y \in C_x} A_y$ is connected by Ex. 4.12, since $x \in A_y$ for any $y \in C_x$. Also clearly

$$C_x \subseteq \bigcup_{y \in C_x} A_y.$$

On the other hand if $z \in \bigcup_{y \in C_x} A_y$, then $z \in A_y$ for some y and since $x \in A_y$, and A_y is connected, x and z belong to some connected subset of M, whence $z \in C_x$. Hence

$$\bigcup_{y \in C_x} A_y \subseteq C_x, \quad \text{whence } C_x = \bigcup_{y \in C_x} A_y,$$

and C_x is connected.

(2) Let $A \subseteq M$, $x \in A$, and let A be connected. For any $z \in A$, x and z lie in a connected subset of M, namely A, hence $z \in C_x$, and $A \subseteq C_x$. ∎

Until Theorem 4.11 was established we had a rather cumbersome mode of talking about components of a set. We had to refer to the component of x in M. Now that we know that components are merely maximal connected subsets of M, we no longer have to refer to "the component of x in M" but can simply talk about components of M. In this connection see also Ex. 4.18.

4.12. Theorem. *Let X be a space, M a closed subset of X, $x \in M$, then C_x, the component of x in M, is also closed in X.*

Proof. Since $C_x \subseteq \overline{C_x}$, $\overline{C_x}$ is connected by 4.9. Since M is closed, $C_x \subseteq M$ implies $\overline{C_x} \subseteq \overline{M} = M$. Finally by 4.11(2) $C_x = \overline{C_x}$, and C_x is closed. ∎

EXAMPLES (AND EXERCISES)

4.16. If X is a topological space, $M \subseteq X$ is open, then there is little one can say about the components of M beyond what has already been said about components in general. In particular we can conclude nothing about components being either open or closed or neither. Consider the space

$$X = \left\{(x, y) \mid 0 \leq x \leq 1, y = \frac{1}{n}x, n = 1, 2, \ldots, \text{ or } y = 0\right\}$$

with the relative topology inherited from the real plane. Let $M = X - \{(0, 0)\}$, and let $z = (1, 0)$, then M is open and the component of z in M,

$$C_z = \{(x, y) \mid 0 < x \leq 1, y = 0\}.$$

Is C_z closed in X? Surely not, since $(0, 0) \notin C_z$. Is C_z open in X? Surely not, since any neighborhood of z contains points of the form $(1, 1/n)$ for n sufficiently large. For a picture of X see Fig. 4.1.

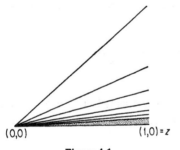

Figure 4.1

4.17. Show that if X is a space, $x \in M \subseteq X$, and if $y \in C_x$, then $C_x = C_y$.

4.18. Let X be a space, $M \subseteq X$, $x, y \in M$. Define $x \sim y$ if and only if there exists a connected subset N of M such that $x, y \in N$. Show that "∼" is an equivalence relation in M, and show further that the equivalence classes induced by "∼" in M are exactly the components of M.

§4 Local-Connectedness

We would like to introduce a property of spaces which would allow us to draw some conclusions about components of open sets. By analogy to closed sets, it would certainly be pleasant if in some spaces, at least, components of open sets were open. Let us proceed boldly, and merely give spaces that have this desired property a new name.

4.13. Definition. *A space X is said to be **locally connected** if and only if components of open sets are open.*

The name "locally connected" for such spaces may seem ill chosen; however, as will shortly appear, it is quite in harmony with our definition of another local property, namely, local compactness. The next theorem exhibits exactly the analogy with Definition 3.29 that we want.

4.14. Theorem. *Let X be a space, then X is locally connected if and only if X has a basis of connected sets, i.e., for each $x \in X$ and each $U \in \mathfrak{U}_x$ there exists V, open, connected, $V \in \mathfrak{U}_x$ such that $V \subseteq U$.*

Proof. Let X be locally connected. Define

$$\mathfrak{B} = \{C \mid C \subseteq X, C \text{ connected and open}\}.$$

We show that $\mathfrak{B}$ is a basis. Clearly $\mathfrak{B} \subseteq \mathfrak{O}$, further let $x \in X$, $U \in \mathfrak{U}_x$, then there is an $O \in \mathfrak{O}$ such that $x \in O \subseteq U$. Let $V = C_x$, the component of x in O, then V is open by 4.13 and V is connected by 4.11(1), whence $V \in \mathfrak{B}$. Also $x \in V \subseteq O \subseteq U$, thus $\mathfrak{B}$ is a basis.

Conversely suppose $\mathfrak{B}$ is a basis of connected sets for X. Let $M \subseteq X$, M open, let $x \in M$, and let C_x be the component of x in M. Let $y \in C_x \subseteq M$, then there is a $U \in \mathfrak{B}$ such that $y \in U \subseteq M$. Since $y \in U$ and U is connected, $U \subseteq C_y$, the component of y in M, by 4.11; but by Ex. 4.17, $C_x = C_y$, thus for each $y \in C_x$, there exists $U \in \mathfrak{U}_y \cap \mathfrak{B}$ such that $y \in U \subseteq C_x$, whence C_x is open. ∎

The student should be warned that Definition 4.13 is rather atypical in that, should he look in other texts on topology, he will not find locally connected spaces so defined. Rather he will find them defined as Theorem 4.14 suggests, i.e., so that each neighborhood of each point contains a connected subneighborhood. Once we have proved 4.14, however, it makes no difference how we define "locally connected," and certainly our definition has more intuitive appeal.

EXAMPLES (AND EXERCISES)

4.19. Show that a space X is locally connected if and only if for each $U \in \mathcal{U}_x$, the component C_x of x in U also belongs to $\mathcal{U}_x$.

4.20. Let
$$X = \left\{(x, y) \mid x = \frac{1}{n}, 0 \leq y \leq 1, \text{ for } n = 1, 2, \ldots\right\}$$
$$\cup \; \{(0, 0), (0, 1)\}$$

have the relative topology inherited from the usual topology of the plane. Show that $\{(0, 0)\}$ and $\{(0, 1)\}$ are components in X. Show further that if $D \subseteq X$ is both open and closed, then either both $\{(0, 0)\}$ and $\{(0, 1)\}$ lie in D or neither of them does.

4.21. Let X be a space which is locally connected and connected.
  (a) Show that if C is a component of an open set $A \subseteq X$, $A \neq X$, then $\text{Fr}\,(C) \subseteq A^c$. (Cf. 1.16.)
  (b) Let M and N be disjoint nonempty closed sets in X. Show that there is a component C of $(M \cup N)^c$ such that $\overline{C} \cap M \neq \emptyset$ and $\overline{C} \cap N \neq \emptyset$.
  (c) Let B be a closed set in X, $B \neq X$, C a component of B, show that $C \cap \overline{B^c} \neq \emptyset$.

4.22. Let $f : X \to Y$ be a closed mapping onto and let X be locally connected. Show that Y is locally connected.

4.23. Show that any open subset of a locally connected space is again locally connected.

4.24. (*Term paper, junior grade*) For each $\alpha \in A$, let X_α be a Hausdorff space.

(a) Show that $\underset{\alpha \in A}{\times} X_\alpha$ is connected if and only if each X_α is connected.

(b) Show that $\underset{\alpha \in A}{\times} X_\alpha$ is locally connected if and only if each X_α is locally connected and each X_α is connected with at most a finite number of exceptions.

[*Hint for* (a). Prove first that if $f : X \to Y$, Y a Hausdorff space, is continuous, $D \subseteq X$ is dense in X, and $f(D) = y_0 \in Y$ is a fixed point of Y, then $f(X) = y_0$. Assume $\underset{\alpha \in A}{\times} X_\alpha = X$ is not connected, then use 4.3 to define $f : X \to R$ so that $f(X) = \{a, b\}$, $a \neq b$, $a, b \in R$. Use the homeomorphism obtained in Ex. 2.22 to show that f is constant (say, $f(x) = a$) on sets of the form X'_β defined in Ex. 2.22. Let

$$X'_{\beta_1 \beta_2 \ldots \beta_n} = \{x \mid x_\alpha = x'_\alpha, \alpha \neq \beta_i, i = 1, 2, \ldots, n, x_{\beta_i} \text{ arbitrary}\}$$

as in Ex. 2.22 and show that for $x \in X'_{\beta_1 \beta_2 \ldots \beta_n}$, $f(x) = a$. Then show that sets of the form $X'_{\beta_1 \beta_2 \ldots \beta_n}$ are dense in X, and finally that $f^{-1}(b) = \emptyset$ to obtain a contradiction.]

[*Hint for* (b). Use (a).]

§ 5 Arcwise Connectedness

We attack finally one more type of connectedness which is of considerable utility in other branches of topology. The motivation here is that some of the spaces that are quite familiar have the property that points may be joined by means of arcs—for example, the real line or the real plane or for that matter any Euclidean space $E^n = \underset{i=1}{\overset{n}{\times}} R_i$, where each R_i is the reals with the usual topology. We generalize this notion via the following:

4.15. Definition. (1) *Let X be a topological space, $A \subseteq X$, and let $I = \{x \mid 0 \leq x \leq 1, x \text{ real}\}$ with the relative topology inherited from the usual topology of the reals. A is said to be an* **arc** *provided there exists a homeomorphism h of I with A. We say that the arc A is from the point $x_0 = h(0)$ to the point $x_1 = h(1)$, or simply from x_0 to x_1.*

(2) Let X be a topological space, then X is said to be **arcwise connected** if for each $x, y \in X$ there is an arc from x to y.

We shall not explore arcwise connected spaces to any great extent, but shall content ourselves with the following two theorems.

4.16. Theorem. *Let X be an arcwise connected space, then X is connected.*

Proof. Let $x \in X$ and for each $y \in X$, let $I_{xy} = h(I)$ be an arc from x to y, then $x \in \bigcap_{y \in X} I_{xy}$ and this intersection is nonempty. Furthermore by 4.4 each I_{xy} is connected, consequently by Ex. 4.12, $\bigcup_{y \in X} I_{xy} = X$ is connected. ∎

4.17. Theorem. *Arcwise connectedness is a topological property.*

Proof. Let $f : X \to Y$ be a homeomorphism and let $y_1, y_2 \in Y$, then there exist $x_1, x_2 \in X$ such that $f(x_1) = y_1, f(x_2) = y_2$. Since X is arcwise connected there exists a homeomorphism $h : I \to X$ such that $h(0) = x_1$ and $h(1) = x_2$. $fh : I \to Y$ is a homeomorphism of I into Y such that $fh(0) = y_1$ and $fh(1) = y_2$, whence Y is arcwise connected. ∎

EXAMPLES (AND EXERCISES)

4.25. The space

$$X = \left\{(x, y) \mid y = \sin \frac{1}{x}, 0 < x \leq \frac{1}{\pi}\right\} \cup \{(0, 0)\}$$

is not arcwise connected, but is connected in the relative topology inherited from the plane.

4.26. Not every arcwise connected space is locally connected; consider the space of Ex. 4.16.

4.27. We now use some of the results obtained so far in this chapter to prove a fixed-point theorem. Prove that if $f : I \to I$ is a mapping where $I = \{x \mid 0 \leq x \leq 1, x \text{ real}\}$, then there is an $x \in I$ such that $f(x) = x$. [*Hint.* If $f(0) = 0$ or $f(1) = 1$ we are finished;

if neither of these happens, $f(0) > 0$ and $f(1) < 1$. In $I \times I$, which is an arcwise connected space (why?), let

$$U = \{(x, y) \mid 1 \geq y > x \geq 0\},$$

let $x_0 = (0, f(0))$ and $x_1 = (1, f(1))$, then

$$x_0 \in U, \quad x_1 \in U^c$$

and $f : I \to I$ defines a path, i.e., a continuous image of $[0, 1]$, from x_0 to x_1. Let $A = f(I)$, then $A \cap U \neq \emptyset$, and $A \cap U^c \neq \emptyset$, thus by Ex. 4.14

$$A \cap \text{Fr}(U) \neq \emptyset.$$

Show that

$$\text{Fr}(U) = \{(x, y) \mid x = y, 0 \leq x \leq 1\},$$

and that if $(x, y) \in A \cap \text{Fr}(U)$, then $f(x) = x$.]

As a closing topic in this chapter we investigate the other side of the coin and look at spaces which are as badly disconnected as possible. To do this we look at the components of the space, and if the space is to be as badly disconnected as possible, we shall want the components to be as small as possible, namely points. We thus make the following:

4.18. Definition. *A space X is said to be **totally disconnected** if and only if all of its components are points.*

There are, of course, many examples of such spaces. Any space with the discrete topology is totally disconnected; thus, for example, the rationals as a subspace of the reals are totally disconnected. Any finite T_1 space is totally disconnected. We have in mind, however, looking at a more interesting example.

EXAMPLES (AND EXERCISES)

Let
$I = \{x \mid 0 \leq x \leq 1, x \text{ real}\},$
$I_1^1 = \{x \mid 1/3 < x < 2/3, x \text{ real}\}, \quad C_1 = I - I_1,$

$$I_2^1 = \{x \mid 1/9 < x < 2/9\}, \quad I_2^2 = \{x \mid 7/9 < x < 8/9\},$$
$$C_2 = C_1 - (I_2^1 \cup I_2^2),$$

$$\vdots$$

$$I_n^1 = \{x \mid 1/3^n < x < 2/3^n\}, \quad I_n^2 = \{x \mid 7/3^n < x < 8/3^n\},$$
$$\ldots, \quad I_n^n = \{x \mid (3^n - 2)/3^n < x < (3^n - 1)/3^n\},$$
$$C_n = C_{n-1} - \left(\bigcup_{i=1}^{n} I_n^i\right), \text{ and finally}$$

$$C = \bigcap_{n=1}^{\infty} C_n.$$

What we are doing is to remove from the unit interval the open middle third, and then at each successive step to remove the open middle third from each interval that is left after the preceding middle thirds have been removed. See Fig. 4.2.

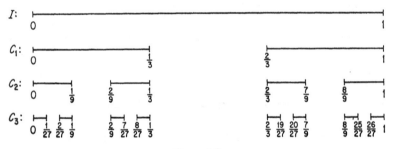

Figure 4.2

There is a great deal known about C, called the Cantor set, or sometimes the Cantor middle third, or ternary set. We mention a few facts here, which we leave to the student to verify.

4.28. If it is agreed to write all numbers x, $0 \leq x \leq 1$ in the scale 3 (i.e., 0, 1, 2 are the only digits which occur) and if all infinite repeating decimals in which eventually an infinite string of zeros preceded by a one occurs are replaced by a decimal in which the one is replaced by a zero in the place before the infinite string of zeros, and this place followed by an infinite string of two's, e.g., .001000 . . . is replaced by .000222 . . . and if similarly .1222 . . .

is replaced by .2000, then the points of C are exactly those decimals written in the scale of 3 in which no one ever occurs.

4.29. From Ex. 4.28 above it is clear that there is a one-to-one correspondence between the class of all functions on

$$Z^+ = \{n \mid n \text{ a positive integer}\}$$

to a set consisting of two objects (namely 0 and 2) and C. The class of such functions, written 2^{Z^+} or $2^{\aleph_0}$ is known to be uncountable, thus C is uncountable.

4.30. C is closed in R.

4.31. C is dense in itself, i.e., for each $x \in C$, x is a limit point of $C - \{x\}$.

4.32. C is nowhere dense in I, i.e., if $x, y \in C$, $x < y$, then there is an interval $J \subseteq I - C$, $J = (a, b)$, $x < a < b < y$.

4.33. C is totally disconnected. [*Hint.* Use Ex. 4.32 above.]

4.34. C is compact. [*Hint.* Use the fact that I is compact, and Ex. 4.30 above.]

It is, in fact, the case that any compact, totally disconnected, dense in itself, metric (to be defined in the next chapter) space is homeomorphic to the Cantor set; however, we make no attempt to prove this here. The interested student is referred to more advanced works on topology for a proof of this, e.g., Hocking and Young, *Topology*, p. 100. The reason these last remarks are made about the Cantor set is that the stated properties are a complete set of invariants for the Cantor set; that is to say, if we have a topological space which possesses the given properties, namely is compact, totally disconnected, dense in itself, and metric, then it is topologically indistinguishable from the Cantor set. Of course, it may be distinguishable from the Cantor set in some other, nontopological way, say in that its points are blue, or are named Sam; but as topologists, this is of no concern to us.

4.35. A point x in a connected topological space X is called a **cut point** of X, if $X - \{x\}$ is not connected. Prove the following:

(a) If X is homeomorphic to Y under f and x is a cut point of X, then $f(x)$ is a cut point of Y.
(b) Every point of the real line is a cut point.
(c) No point of the real plane is a cut point.
(d) The real line and the real plane are not homeomorphic.

CHAPTER 5

METRIC SPACES

§ 1 Definitions

We apply ourselves now to a rather special sort of topological space, one in which there is defined a distance function, so that we can say what the distance between points is. In a sense these spaces, so-called metric spaces, are rather special, since, as will turn out in the sequel, they will enjoy properties of the sorts we have already discussed, but under less restrictive hypotheses than more general spaces. On the other hand, metric spaces are still quite general, since all the common spaces of analysis are metric spaces. We define a metric space in two stages; first we make the following:

5.1. Definition. *Let X be a set, let $\rho : X \times X \to R$ be a function (not necessarily continuous) of $X \times X$ into R, the reals, such that*

(1) $\rho(x, y) \geq 0$ and $\rho(x, y) = 0$ if and only if $x = y$,
(2) $\rho(x, y) = \rho(y, x)$,
(3) $\rho(x, y) + \rho(y, z) \geq \rho(x, z)$,

*then (X, ρ) is called a **metric set**, and ρ is called a **metric** for X.*

Unfortunately 5.1 is so general that every set is a metric set, for define $\rho(x, y) = 1$ if $x \neq y$, and $\rho(x, y) = 0$ if $x = y$. It is a simple verification that with this definition of ρ, any set X is a metric set. What we are really after is to have the metric generate the topology for X. To that end we make the following:

5.2. Definition. *Let (X, ρ) be a metric set, define for $x \in X$*

$$S_\epsilon(x) = \{y \mid \rho(x, y) < \epsilon \text{ for } y \in X, \epsilon > 0 \text{ and real}\}$$

and define

$$\mathcal{B} = \{S_\epsilon(x) \mid x \in X, \epsilon > 0\}.$$

Then X with $\mathcal{B}$ as basis is called a **metric space**. The topology so generated is called a **metric topology** generated by ρ.

Now only the rankest of amateurs at mathematics tries to prove definitions; however, the above definition makes some assertions which must be verified. In particular, it is asserted that $\mathcal{B}$ is a basis, and this is perhaps not so evident without some proof. We deal with this minor matter in the following:

5.3. Theorem. *The set* $\mathcal{B} = \{S_\epsilon(x) \mid x \in X, \epsilon > 0\}$ *is a basis for some topology for the metric set* (X, ρ).

Proof. We shall apply 1.29. Since $x \in S_\epsilon(x)$ for each $x \in X$ clearly we have

$$\bigcup_{S_\epsilon(x) \in \mathcal{B}} S_\epsilon(x) = X.$$

Now let $x \in X$ and let $S_{\epsilon_1}(x_1), S_{\epsilon_2}(x_2) \in \mathcal{B}$ so that

$$x \in S_{\epsilon_1}(x_1) \cap S_{\epsilon_2}(x_2).$$

Let $\rho(x, x_1) = d_1 < \epsilon_1$ and $\rho(x, x_2) = d_2 < \epsilon_2$, and let

$$\epsilon = \min(\epsilon_2 - d_2, \epsilon_1 - d_1).$$

We consider $S_\epsilon(x)$; let $y \in S_\epsilon(x)$, then $\rho(y, x) < \epsilon$. Now

$$\rho(y, x_2) \leq \rho(y, x) + \rho(x, x_2) < \epsilon + d_2$$
$$\leq (\epsilon_2 - d_2) + d_2 = \epsilon_2,$$

so that $y \in S_{\epsilon_2}(x_2)$. Also

$$\rho(y, x_1) \leq \rho(y, x) + \rho(x, x_1) < \epsilon + d_1$$
$$\leq (\epsilon_1 - d_1) + d_1 = \epsilon_1,$$

so that $y \in S_{\epsilon_1}(x_1)$. Thus we have found $S_\epsilon(x) \in \mathcal{B}$ so that

$$S_\epsilon(x) \subseteq S_{\epsilon_1}(x_1) \cap S_{\epsilon_2}(x_2),$$

and by 1.29, $\mathcal{B}$ is a basis. ∎

SEC. 5.1   METRIC SPACES   117

NOTE: The $S_\epsilon(x)$ defined in 5.2 and used so extensively in 5.3 are called spherical ϵ-neighborhoods of x.

The sort of computation with the metric that appears in 5.3 is fairly typical of what goes on in metric spaces. The student should familiarize himself with the techniques of such computations, and should draw suitable pictures to assist him. For example, the choice of ϵ in the proof of 5.3 is motivated by the situation exhibited in Fig. 5.1.

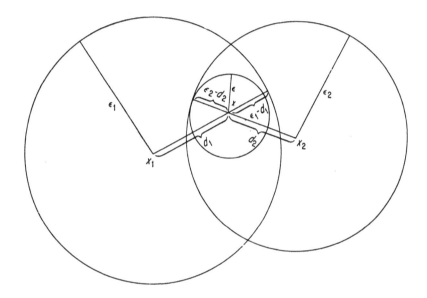

Figure 5.1

Now, of course, it can happen that a space X is already given us, and we may wish to know if it is possible to define a metric, ρ, so that the topology generated by the metric, using 5.2, is in fact the same as the original topology. We define a space with this desirable property in the following:

5.4. Definition. *Let X be a topological space with topology $\mathfrak{T}$. If it is possible to define a metric ρ so that the metric topology generated by ρ coincides with $\mathfrak{T}$, then X is said to be a* **metrizable space.**

EXAMPLES (AND EXERCISES)

5.1. The real line, R, with $\rho(x, y) = |x - y|$ is a metric space, and the metric so defined generates a topology which is the same as the usual topology for R.

5.2. The real plane with

$$\rho((x_1, y_1), (x_2, y_2)) = [(x_1 - x_2)^2 + (y_1 - y_2)^2]^{1/2}$$

is a metric space, and the metric so defined generates a topology which is the same as the usual topology for the real plane.

5.3. Let (X, ρ) and (X', ρ') be metric spaces and let f be a relation on X to X' [i.e., f is a (possibly) multivalued function with domain X and range X']. Further, for each $x, y \in X$, let $\rho(x, y) = \rho'(f(x), f(y))$ where $f(x)$ (or $f(y)$) represents any one of the values of f at x (or y), then f is called an isometry from X to X'. Show that if f is an isometry, then it is a homeomorphism (possibly into X').

5.4. Show that what is suggested by the last three words of Ex. 5.3 cannot occur if f is an isometry of X with itself and X is compact, i.e., show that if $f : X \to X$ is an isometry and X is compact, then f is necessarily onto.

5.5. Let X be a set, and let $\rho(x, y)$ be a real-valued function on $X \times X$ such that

(a) $\rho(x, y) = 0$ if and only if $x = y$.
(b) $\rho(x, y) \leq \rho(x, z) + \rho(y, z)$ for all $x, y, z \in X$.

Then (X, ρ) is a metric set.

NOTE: Before attacking the next three exercises it is useful to observe that in every metric space X, if $A \subseteq X$ and x is a limit point of A, then there exists a sequence $\{x_n\}$ of distinct points of A, such that $\lim_n x_n = x$.

5.6. Let X be a metric space with metric ρ, and let $\varnothing \neq A \subseteq X$. Define the **diameter of A**, $\delta(A) = \sup_{x,y \in A} \rho(x, y)$, i.e., the least upper bound of distances between points of A. Show that

(a) If $A \subseteq B$, then $\delta(A) \leq \delta(B)$.
(b) $\delta(A) = \delta(\overline{A})$.
(c) If A is compact, then there exist $x, y \in A$ such that $\delta(A) = \rho(x, y)$.
(d) There exist sets A (obviously not compact) such that for all $x, y \in A$, $\rho(x, y) < \delta(A)$.

5.7. Let X be a metric space with metric ρ, let $x \in X$, and let $\emptyset \neq A \subseteq X$; define $\rho(x, A) = \inf_{y \in A} \rho(x, y)$. $\rho(x, A)$ is called the **distance from the point x to the set A**. Show that
(a) If A is compact, there exists a point $y \in A$ such that $\rho(x, A) = \rho(x, y)$.
(b) There exist sets A (obviously not compact) such that for all $y \in A$, $\rho(x, A) < \rho(x, y)$.
(c) $\overline{A} = \{x \mid \rho(x, A) = 0\}$.

5.8. Let X be a metric space with metric ρ, let $\emptyset \neq A \subseteq X$, $\emptyset \neq B \subseteq X$; define the **distance from A to B**,

$$\rho(A, B) = \inf_{x \in A, y \in B} \rho(x, y).$$

Show that

(a) $\rho(A, B) = \rho(\overline{A}, \overline{B})$.
(b) If A and B are compact, then there exist $x \in A$, $y \in B$ such that $\rho(A, B) = \rho(x, y)$.
(c) There exist closed sets, A and B (obviously not both compact), such that for all $x \in A$, $y \in B$

$$\rho(A, B) < \rho(x, y).$$

(d) For any sets $A, B, C \subseteq X$,

$$\rho(A, C) \leq \rho(A, B) + \rho(A \cup B, C) + \delta(B).$$

(e) For any sets $A, B, C \subseteq X$,

$$\rho(A, C) \leq \rho(A, B) + \rho(B, C) + \delta(B).$$

Give an example in which $\delta(B) \neq 0$ in which equality holds.

§ 2 Some Properties of Metric Spaces

One of the more interesting problems is to try to decide what sorts of spaces are metrizable. Before we tackle this problem, however, let us explore some of the properties of metric spaces. Metric spaces have a fairly strong structure, and this is demonstrated by the following theorems.

5.5. Theorem. *Every metric space is a Hausdorff space.*

Proof. Let $x, y \in X$, with $x \neq y$, then $\rho(x, y) = d > 0$; let $\epsilon = d/2$ and let $U = S_\epsilon(x)$, $V = S_\epsilon(y)$, then U and V are open sets, and we need only show $U \cap V = \emptyset$.

Suppose in fact that $U \cap V \neq \emptyset$, then there is a $z \in U \cap V$ and since $x \in U$, $\rho(z, x) < \epsilon$. Similarly $\rho(y, z) < \epsilon$. Thus

$$\rho(x, y) \leq \rho(x, z) + \rho(y, z) < 2\epsilon = d,$$

and $d = \rho(x, y) < d$. This palpable contradiction shows that

$$U \cap V = \emptyset,$$

whence X is Hausdorff. ∎

5.6. Theorem. *Every metric space is a normal space.*

Proof. We already know that if X is metric, then it is a Hausdorff space, hence T_1. We need thus only verify that if A and B are closed disjoint subsets of X, then there exist open sets U, V such that $A \subseteq U$, $B \subseteq V$, and $U \cap V = \emptyset$.

Now for each $a \in A$, a is not a limit point of B, for if it were we should have $a \in B$, since B is closed; and then $a \in A \cap B = \emptyset$ and this is clearly impossible. Thus for each $a \in A$, there exists an $\epsilon_a > 0$ so that $S_{\epsilon_a}(a) \cap B = \emptyset$. Let $U_a = S_{\epsilon_a/2}(a)$. Similarly for each $b \in B$, there is an $\epsilon_b > 0$ so that $S_{\epsilon_b}(b) \cap A = \emptyset$. Let $V_b = S_{\epsilon_b/2}(b)$. Finally let

$$U = \bigcup_{a \in A} U_a, \qquad V = \bigcup_{b \in B} V_b,$$

then U and V are open.

SEC. 5.2   METRIC SPACES   121

We show now that $U \cap V = \emptyset$. Suppose in fact that $U \cap V \neq \emptyset$, and let $x \in U \cap V$, then $x \in U_a$ for some a and $x \in V_b$ for some b, whence $\rho(x, a) < \epsilon_a/2$, $\rho(x, b) < \epsilon_b/2$, and

$$\rho(a, b) \leq \rho(a, x) + \rho(x, b) < \tfrac{1}{2}(\epsilon_a + \epsilon_b).$$

If $\epsilon_a \leq \epsilon_b$, then

$$\rho(a, b) < \tfrac{1}{2}(\epsilon_a + \epsilon_b) \leq \epsilon_b \quad \text{and} \quad a \in S_{\epsilon_b}(b),$$

so that $S_{\epsilon_b}(b) \cap A \neq \emptyset$, contrary to the choice of ϵ_b. On the other hand if $\epsilon_b \leq \epsilon_a$, then

$$\rho(a, b) < \tfrac{1}{2}(\epsilon_a + \epsilon_b) \leq \epsilon_a \quad \text{and} \quad b \in S_{\epsilon_a}(a),$$

so that $S_{\epsilon_a}(a) \cap B \neq \emptyset$, contrary to the choice of ϵ_a. It must be the case that $U \cap V = \emptyset$, and hence that X is normal. ∎

Further exploration of the properties of metric spaces shows that notions which were heretofore distinct become equivalent in metric spaces. The following few theorems demonstrate this.

5.7. Theorem. *Let X be a metric space, then X is separable if and only if X is second countable.*

Proof. Let X be separable, and let

$$D = \{x_i \mid i = 1, 2, \ldots\}$$

be a countable dense subset, let

$$\mathcal{B} = \{S_r(x_i) \mid x_i \in D, r \text{ rational}, r > 0\},$$

then $\mathcal{B}$ is a countable family of open sets, being the countable union of families of the form

$$\{S_r(x_i) \mid x_i \text{ fixed}, r \text{ rational}, r > 0\}$$

which are countable since the rationals are countable. We show now that $\mathcal{B}$ is a basis.

Let $x \in X$, $U \in \mathcal{U}_x$, then there exists $\epsilon > 0$ such that $S_\epsilon(x) \subseteq U$. If $x = x_i$ for some i, select r, rational, $0 < r < \epsilon$, then

$$x \in S_r(x) \subseteq S_\epsilon(x) \subseteq U.$$

If $x \neq x_i$ for any i, then x is a limit point of D and there exists an i such that $x_i \in S_{\epsilon/3}(x)$. Select r, rational, so that $\epsilon/3 < r < \epsilon/2$, then since $\rho(x, x_i) < \epsilon/3 < r$, $x \in S_r(x_i)$. Also if $y \in S_r(x_i)$, then

$$\rho(x, y) \leq \rho(x, x_i) + \rho(x_i, y) < \frac{\epsilon}{3} + \frac{\epsilon}{2} = \frac{5\epsilon}{6} < \epsilon$$

so that $y \in S_\epsilon(x)$ and we have

$$x \in S_r(x_i) \subseteq S_\epsilon(x) \subseteq U.$$

In any case we have found $S_r(x_i) \in \mathfrak{B}$ such that $x \in S_r(x_i) \subseteq U$, whence $\mathfrak{B}$ is a basis.

The converse is settled by 1.32. ∎

5.8. Lemma. *Let X be a countably compact metric space, and let $\epsilon > 0$, then there exists a finite set F_ϵ such that $X = \bigcup_{x \in F_\epsilon} S_\epsilon(x)$.*

Proof. Suppose the theorem false. Select $x_1 \in X$, then $X \neq S_\epsilon(x_1)$, and we may choose $x_2 \in X - S_\epsilon(x_1)$. Note that $\rho(x_1, x_2) \geq \epsilon$. Once again

$$X \neq \bigcup_{i=1}^{2} S_\epsilon(x_i),$$

and we proceed similarly; at the nth step,

$$X \neq \bigcup_{i=1}^{n} S_\epsilon(x_i)$$

and we may select $x_{n+1} \in X - \bigcup_{i=1}^{n} S_\epsilon(x_i)$ and we observe that

$$\rho(x_i, x_j) \geq \epsilon \quad \text{for } i \neq j, \quad 1 \leq i, j \leq n + 1.$$

Consider the sequence $\{x_n\}$ thus constructed. It is an infinite subset of X, since $x_i \neq x_j$ for $i \neq j$, thus since X is countably compact $\{x_n\}$ has a limit point $x_0 \in X$. There then exist infinitely many points of $\{x_n\}$ in $S_{\epsilon/2}(x_0)$, hence a fortiori at least two distinct such points, say x_m and x_n. Then

$$\rho(x_m, x_n) \leq \rho(x_m, x_0) + \rho(x_n, x_0) < \frac{\epsilon}{2} + \frac{\epsilon}{2} = \epsilon,$$

and this contradicts
$$\rho(x_m, x_n) \geq \epsilon.$$
This contradiction proves the lemma. ∎

5.9. Lemma. *Every countably compact metric space is second countable.*

Proof. For each positive integer n, let $F_{1/n}$ be the finite set, whose existence is guaranteed by 5.8 such that
$$\bigcup_{x \in F_{1/n}} S_{1/n}(x) = X.$$

Let $D = \bigcup_{n=1}^{\infty} F_{1/n}$, then D is the countable union of finite sets and is thus again countable. We show $\overline{D} = X$. Let $x \in X - D$ and let $\epsilon > 0$. Select $n > 0$ so that $1/n < \epsilon$, then since
$$X = \bigcup_{x \in F_{1/n}} S_{1/n}(x),$$
there exists $y \in F_{1/n} \subseteq D$ so that $x \in S_{1/n}(y)$, whence
$$\rho(x, y) < \frac{1}{n} < \epsilon \quad \text{and} \quad y \in S_\epsilon(x).$$

There thus exists for each $\epsilon > 0$ a $y \in D \cap S_\epsilon(x)$, consequently $x \in \overline{D}$, and $\overline{D} = X$. X is thus separable, and therefore by 5.7 second countable. ∎

5.10. Theorem. *A metric space is compact if and only if it is countably compact.*

Proof. Compactness implies countable compactness by 3.26. Conversely if X is countably compact, then by 5.9, X is second countable. Since X is metric it is also Hausdorff by 5.5 and is consequently a T_1 space. The desired conclusion now follows from 3.28. ∎

EXAMPLES (AND EXERCISES)

5.9. Go back and look at Definition 3.22. Now prove that a T_1 space X is completely normal if and only if, whenever A and B are two separated subsets of X, then there exist disjoint open sets, U and V, such that $A \subseteq U$, $B \subseteq V$.

5.10. Show that every metric space is completely normal. [*Hint.* Show first that every subspace of a metric space is metric, i.e., if $A \subseteq X$, then the metric is compatible with the relative topology.]

5.11. Show that if X is a metric space, then the mapping $\rho : X \times X \to R$ defined by $\rho[(x, y)] = \rho(x, y)$, i.e., the distance from x to y, is a continuous function.

5.12. Show that if X is a connected metric space, then for each $x \in X$, the function $f_x(y) = \rho(x, y)$ has the intermediate value property [i.e., if $f_x(y_1) = a < c < b = f_x(y_2)$, then there exists $y \in X$ such that $f_x(y) = c$].

5.13. Show that there exist metric spaces X which are not connected yet for which for each $x \in X$, $f_x(y) = \rho(x, y)$ has the intermediate value property.

5.14. Show that any subspace of a separable metric space is again a separable metric space. [*Hint.* Use 5.7 and Ex. 1.72.] Show further that if X is not metric this result may fail. [*Hint.* Consider the space J of Ex. 1.74, and let $P = \text{Fr}(J)$ in E be the subspace.]

5.15. Let X be a metric space with metric ρ. Define $\rho(A, B)$ as in Ex. 5.8. Show that in general $\rho(A, B)$ is not a metric for the set of all subsets of X.

Now define $S_\epsilon(A) = \bigcup_{x \in A} S_\epsilon(x)$, and define

$$\rho_h(A, B) = \inf_{\epsilon > 0} \{\epsilon \mid B \subseteq S_\epsilon(A) \text{ and } A \subseteq S_\epsilon(B)\}.$$

Let X be a metric space, and let $X^* = \{C \mid C \subseteq X, C \text{ closed}\}$, then ρ_h is a metric for X^*. The metric ρ_h is usually called the **Hausdorff metric** for the space X^*.

5.16. Show that any metric space X is homeomorphic to a metric space X' with metric ρ' for which $\rho'(x', y') < 1$ for all $x', y' \in X'$. [*Hint.* Let the points of X' be the same as the points of X, and define $\rho'(x, y) = \rho(x, y)/[1 + \rho(x, y)]$, where ρ is the metric in X. Then show that ρ' is a metric, $\rho'(x, y) < 1$ for all $x, y \in X$, and that the mapping $f(x) = x$ is a homeomorphism.]

§3 Metrization Theorems

The object of the following sequence of theorems and lemmas is to establish that every regular second countable space is metrizable in the sense of Definition 5.4. We require a good bit of machinery before we are ready to prove this result, and it is the purpose of the following lemmas to assemble the appropriate hardware.

First a definition:

5.11. Definition. *Let*

$$\mathcal{H} = \left\{ y \mid y = \{y_n\},\ y_n\ \text{real for each}\ n\ \text{such that}\ \sum_{n=1}^{\infty} y_n^2 < \infty \right\},$$

i.e., $\mathcal{H}$ is the collection of all sequences of real numbers such that the series formed from the squares of the terms of the sequence is a convergent series. Define, for $x, y \in \mathcal{H}$,

$$\rho(x, y) = \left[\sum_{i=1}^{\infty} (x_i - y_i)^2 \right]^{1/2},$$

then the resulting metric space is called **Hilbert space.**

EXAMPLES (AND EXERCISES)

5.17. Show that ρ is a metric for $\mathcal{H}$.

5.18. Show that the subspace of $\mathcal{H}$ defined by

$$E^1 = \{x \mid x \in \mathcal{H},\ x = \{x_n\},\ x_n = 0\ \text{for}\ n > 1\}$$

is homeomorphic to the real line with the usual topology.

5.19. More generally $E^n \subseteq \mathcal{H}$, defined by

$$E^n = \{x \mid x \in \mathcal{H},\ x = \{x_i\},\ x_i = 0\ \text{for}\ i > n\}$$

is homeomorphic to Euclidean n-space, i.e., to $\underset{i=1}{\overset{n}{\times}} R_i$, where each

R_i is the reals with the usual topology.

5.20. The subspace $\mathcal{H}'$ of $\mathcal{H}$ defined by

$$\mathcal{H}' = \{x \mid x \in \mathcal{H}, x = \{x_n\}, 0 \leq x_n \leq 1/n \text{ for each } n\}$$

is called the **Hilbert cube** (or Hilbert parallelotrope). Let $I_n = [0, 1]$ for each n, i.e., the unit interval with the relative topology inherited from the reals, and let $I^\omega = \underset{n=1}{\overset{\infty}{\times}} I_n$, then $\mathcal{H}'$ is homeomorphic to I^ω.

5.21. Show that the Hilbert cube is compact.

5.22. Let

$$A = \{x \mid x \in \mathcal{H}, x_i = \delta_j^i, j = 1, 2, \ldots\},$$

where δ_j^i, the **Kronecker delta,** is defined by

$$\delta_j^i = 0 \quad \text{if} \quad i \neq j, \qquad \delta_j^i = 1 \quad \text{if} \quad i = j,$$

i.e., A is the set of unit points on the "coordinate axes" of $\mathcal{H}$. Show that A is closed but not compact.

5.23. Is the unit sphere

$$S = \left\{ x \mid \sum_{i=1}^{\infty} x_i^2 = 1, \text{ where } x = \{x_i\} \right\} \subseteq \mathcal{H}$$

compact? [*Hint.* Use Ex. 5.22 above.]

Our plan of attack on the metrization theorem is to show that every second countable regular space is homeomorphic to a subset of Hilbert space (in fact of the Hilbert cube), and is thus metrizable, since all we need do is use the metric of the subspace of $\mathcal{H}'$ to which the given space is homeomorphic to define a compatible metric. Now in order to define the appropriate mapping of our space X into $\mathcal{H}'$, we need to specify the terms, y_n, of the sequence $y \in \mathcal{H}'$ which is to be the image point of some preselected point $x \in X$. We thus need some device for associating a sequence of real numbers $\{y_n \mid 0 \leq y_n \leq 1/n\}$ with each point of our space.

We wish to exploit the regularity of X, and specifically to make use of 3.18, which tells us that for each open set, consequently for each basic neighborhood B_i of a point $x \in X$, there is another open set, which we may choose as a second basic neighborhood, B_j, such that $x \in B_j \subseteq \overline{B_j} \subseteq B_i$. Then $\overline{B_j}$ and $X - B_i$ will be disjoint closed sets. If we consider all pairs of basic neighborhoods, (B_j, B_i) of X such that $\overline{B_j} \subseteq B_i$, which set of pairs is countable (by the second countability of X), and if we can associate with each such pair a real-valued function λ_i such that $0 \leq \lambda_i(x) \leq 1$, we would at least be partially on our way. In order to accomplish this we prove first the following:

5.12. Lemma (Urysohn). *Let X be a normal space, A, B closed disjoint subsets of X, then there exists a mapping $f : X \to I$, $I = [0, 1]$ such that $f(A) = 0, f(B) = 1$.*

NOTE: It may seem strange that we start with a normal space, since we are interested in second countable regular spaces; however, we know that second countable regular spaces are normal by 3.24 and the exercises following it.

Proof. Stage 1. Let $X - B = G_1$, an open set, since B is closed, then $A \subseteq G_1$ since $A \cap B = \varnothing$. By 3.19, there exists $G_{1/2}$ open such that $A \subseteq G_{1/2} \subseteq \overline{G}_{1/2} \subseteq G_1$.

Stage 2. Again by 3.19, there exist $G_{1/4}$ and $G_{3/4}$ open such that

$$A \subseteq G_{1/4} \subseteq \overline{G}_{1/4} \subseteq G_{1/2} \subseteq \overline{G}_{1/2} \subseteq G_{3/4} \subseteq \overline{G}_{3/4} \subseteq G_1.$$

Stage 3. Once again by 3.19 there exist $G_{1/8}$, $G_{3/8}$, $G_{5/8}$, and $G_{7/8}$ open such that

$$A \subseteq G_{1/8} \subseteq \overline{G}_{1/8} \subseteq G_{1/4} \subseteq \overline{G}_{1/4} \subseteq G_{3/8} \subseteq \overline{G}_{3/8} \subseteq G_{1/2} \subseteq \overline{G}_{1/2} \subseteq G_{5/8}$$
$$\subseteq \overline{G}_{5/8} \subseteq G_{3/4} \subseteq \overline{G}_{3/4} \subseteq G_{7/8} \subseteq \overline{G}_{7/8} \subseteq G_1$$

and so forth up to

Stage N. By 3.19 for each odd integer $2i - 1$, $1 \leq 2i - 1 \leq 2^N - 1$, there exists an open set $G_{(2i-1)/2^N}$ such that $A \subseteq G_{1/2^N}$ and

$$\overline{G}_{(2i-2)/2^N} \subseteq G_{(2i-1)/2^N} \subseteq \overline{G}_{(2i-1)/2^N} \subseteq G_{2i/2^N}$$

for each i.

By induction we construct for each dyadic fraction, t, between 0 and 1, i.e., for each fraction whose denominator is 2^n, $n \geq 0$, an open set G_t such that if t and t' are two dyadic fractions then $t < t'$ if and only if $\bar{G}_t \subseteq G_{t'}$.

Now for $x \in X$, define

$$f(x) = \inf_{x \in G_t} t \quad \text{for } x \notin B$$
$$= 1 \quad \text{for } x \in B.$$

Observe that $A \subseteq G_t$ for all t, thus $f(x) = 0$ if $x \in A$, and also note that $0 \leq f(x) \leq 1$.

We are left with the task of showing f continuous. Let us examine the structure of $f^{-1}([0, y))$ for $0 < y \leq 1$. Now $f(x) \in [0, y)$ provided $0 \leq f(x) < y$, and since the dyadic fractions are dense in $[0, 1]$ there exists a dyadic fraction t_0 such that

$$f(x) = \inf_{x \in G_t} t < t_0 < y.$$

Consequently $x \in G_{t_0}$. On the other hand if $t_0 < y$ and $x \in G_{t_0}$,

$$f(x) = \inf_{x \in G_t} t < t_0 < y \quad \text{and} \quad f(x) \in [0, y).$$

We thus see that $f^{-1}([0, y)) = \bigcup_{t<y} G_t$. Since G_t is open for each t, $f^{-1}([0, y))$ is open for each y.

By a similar argument it is clear that

$$f^{-1}((y, 1]) = \bigcup_{t>y} (X - \bar{G}_t)$$

where $0 \leq y < 1$. Since $G_t \subseteq \bar{G}_t$, $X - G_t \supseteq X - \bar{G}_t$ for each t. Thus

$$\bigcup_{t>y} (X - G_t) \supseteq \bigcup_{t>y} (X - \bar{G}_t).$$

However, if $x \in \bigcup_{t>y} (X - G_t)$, then there exists a $t > y$ such that $x \in X - G_t$, and again by the density of the dyadic fractions in $[0, 1]$ we may select t', a dyadic fraction, such that $t > t' > y$, then $\bar{G}_{t'} \subseteq G_t$, and $X - \bar{G}_{t'} \supseteq X - G_t$, so that $x \in X - \bar{G}_{t'}$ for some $t' > y$, whence

$$x \in \bigcup_{t>y} (X - \bar{G}_t).$$

Thus
$$\bigcup_{t>y} (X - G_t) \subseteq \bigcup_{t>y} (X - \overline{G}_t)$$
and finally
$$\bigcup_{t>y} (X - G_t) = \bigcup_{t>y} (X - \overline{G}_t).$$

Consequently, since each $\overline{G}_t$ is closed, each $X - \overline{G}_t$ is therefore open, and we see that $f^{-1}((y, 1]) = \bigcup_{t>y} (X - \overline{G}_t)$ is open.

Now let U be some open set in $[0, 1]$, such that $f(x) \in U$, then there exists a basic set V in $[0, 1]$ such that $f(x) \in V \subseteq U$, and the basic set V in the relative topology of $[0, 1]$ has one of the following forms:

(1) $[0, y)$, $0 < y < 1$.
(2) $(y, 1]$, $0 < y < 1$.
(3) (y_1, y_2), $0 < y_1 < y_2 < 1$.
(4) $[0, 1]$.

If $V = [0, y)$, then $f^{-1}(V)$ is open by what we have proved above.
If $V = (y, 1]$, then $f^{-1}(V)$ is open by what we have proved above, also.
If $V = (y_1, y_2)$, then $V = V_1 \cap V_2$, where $V_1 = [0, y_2)$, $V_1 = (y_1, 1]$, and $f^{-1}(V) = f^{-1}(V_1) \cap f^{-1}(V_2)$ is open as the intersection of open sets.
Finally if $V = [0, 1]$, $f^{-1}(V) = X$. In any case then $f^{-1}(V)$ is open for any basic set of $[0, 1]$, whence by Ex. 2.12 f is continuous. ∎

We are now in position, having assembled all the heavy hardware, to attack the principal problem.

5.13. Theorem. *Every second countable regular space is homeomorphic to a subset of the Hilbert cube.*

Proof. Let $\mathcal{B} = \{B_i \mid i = 1, 2, \ldots\}$ be the countable basis. We observe by 3.24 and the exercises following that our space is normal, thus there exist pairs of elements of $\mathcal{B}$ such that $\overline{B}_i \subseteq B_j$. Since $\mathcal{B}$ is countable, the collection of all such pairs is again countable; let us call it
$$\mathcal{P} = \{P_n \mid n = 1, 2, \ldots\}$$
where $P_n = (B_i^n, B_j^n)$ and $\overline{B}_i^n \subseteq B_j^n$. Now since $\overline{B}_i^n \cap (X - B_j^n) = \varnothing$,

and both $\overline{B_i^n}$ and $X - B_j^n$ are closed, we may define, by 5.12, a mapping $f_n : X \to I = [0, 1]$ such that $f_n(\overline{B_i^n}) = 0$, $f_n(X - B_j^n) = 1$. Finally define $f : X \to \mathcal{K}'$, the Hilbert cube, by

$$f(x) = \{f_n(x)/n \mid n = 1, 2, \ldots\}.$$

Since for each x, $0 \leq f_n(x) \leq 1$, $f(x) \in \mathcal{K}'$.

First we show f is one-to-one. Let $x \neq y$, then since X is Hausdorff by 3.17, there exist open sets, which we may choose as basic sets, B, B' such that $x \in B$, $y \in B'$, and $B \cap B' = \emptyset$. Further, since X is normal we can find $B'' \in \mathcal{B}$, such that $x \in B'' \subseteq \overline{B''} \subseteq B$, then $x \in \overline{B''}$, $y \in X - B$, and the pair $(B'', B) \in \mathcal{P}$, i.e., for some n, $(B'', B) = (B_i^n, B_j^n)$. Thus

$$f_n(x) = f_n(\overline{B_i^n}) = f_n(\overline{B''}) = 0,$$

while

$$f_n(y) = f_n(X - B_j^n) = f_n(X - B) = 1.$$

Thus $f(x) \neq f(y)$, since $f(x)$ differs from $f(y)$ at the nth place.

Now we prove f is continuous. Let $x \in X$, and let $\epsilon > 0$. We wish to construct $U \in \mathcal{U}_x$ such that for any $y \in U$,

$$\rho(f(x), f(y)) < \epsilon \text{ in } \mathcal{K}'.$$

First, since for any point $y \in X$, $0 \leq f_n(y) \leq 1$, we have that

$$|f_n(x) - f_n(y)|^2 \leq 1.$$

Now the infinite series $\sum_{n=1}^{\infty} n^{-2}$ converges, thus for N sufficiently large,

$$\sum_{n=N}^{\infty} n^{-2} < \frac{\epsilon^2}{2},$$

whence

$$\sum_{n=N}^{\infty} |f_n(x) - f_n(y)|^2 n^{-2} \leq \sum_{n=N}^{\infty} n^{-2} < \frac{\epsilon^2}{2}.$$

Now let $k < N$, then the function $f_k : X \to I$ is continuous, thus there exists a $U_k \in \mathcal{U}_x$ such that $y \in U_k$ implies

$$|f_k(x) - f_k(y)| < \frac{k\epsilon}{(2(N-1))^{1/2}},$$

or

$$\frac{|f_k(x) - f_k(y)|^2}{k^2} < \frac{\epsilon^2}{2(N-1)}.$$

Now let $U = \bigcap_{k=1}^{N-1} U_k$, then if $y \in U$,

$$\sum_{n=1}^{\infty} \frac{|f_n(x) - f_n(y)|^2}{n^2} = \sum_{n=1}^{N-1} \frac{|f_n(x) - f_n(y)|^2}{n^2} + \sum_{n=N}^{\infty} \frac{|f_n(x) - f_n(y)|^2}{n^2}$$

$$< (N-1)\frac{\epsilon^2}{2(N-1)} + \frac{\epsilon^2}{2} = \epsilon^2,$$

and finally $\rho(f(x), f(y)) < \epsilon$. Thus f is continuous.

Finally we must show that f is an open mapping. Let U be open in X, and let $x \in U$, then there exist $B_i, B_j \in \mathcal{B}$ such that

$$x \in B_i \subseteq \overline{B_i} \subseteq B_j \subseteq U$$

by the normality of X and the fact that $\mathcal{B}$ is a basis. Thus the pair $(B_i, B_j) \in \mathcal{P}$, say $(B_i, B_j) = (B_i^n, B_j^n)$. Then

$$f_n(x) = f_n(\overline{B_i^n}) = 0,$$

and since $X - U \subseteq X - B_j^n$,

$$f_n(X - U) = f_n(X - B_j^n) = 1,$$

so that for any $y \in X - U$,

$$\rho(f(x), f(y)) = \left[\sum_{n=1}^{\infty} \frac{|f_n(x) - f_n(y)|^2}{n^2}\right]^{1/2} \geq \left[\frac{|f_n(x) - f_n(y)|^2}{n^2}\right]^{1/2} = \frac{1}{n}$$

Thus if $V = S_{1/n}(f(x)) \subseteq \mathcal{K}'$, $y \in V$ implies

$$\rho(f(x), y) < \frac{1}{n} \quad \text{and} \quad f^{-1}(y) \in U,$$

for if not,

$$f^{-1}(y) \in X - U \quad \text{and} \quad \rho(f(x), f(f^{-1}(y))) \geq \frac{1}{n},$$

a contradiction. Thus $f^{-1}(V) \subseteq U$, and $x \in V \subseteq f(U)$, whence $f(U)$ is open. Consequently we have proved that f is a one-to-one continuous open mapping, and consequently f is a homeomorphism. ∎

5.14. Theorem. *Every second countable space is metrizable if and only if it is regular.*

Proof. If X is second countable and regular, then by 5.13 X is metrizable.

Conversely, if X is second countable and metrizable, X is a metric space with some compatible metric, ρ. Hence by 5.6 X is normal, and by 3.17 X is regular. ∎

EXAMPLES (AND EXERCISES)

5.24. Show that every locally compact second countable Hausdorff space is metrizable. [*Hint.* Compactify the space.]

5.25. Show that if X is T_1 the converse of Urysohn's lemma (5.12) holds.

§ 4 Complete Metric Spaces

Our plan now is to look at another special class of metric spaces, which possess a distinguishing property, completeness, somewhat weaker than compactness. These spaces are of considerable utility in analysis, and in fact the Banach spaces of analysis are essentially complete metric spaces with some further algebraic structure. Consequently to the student with a bent for analysis, complete metric spaces will be of particular interest.

5.15. Definition. *Let X be a metric space, $\{x_n\} \subseteq X$ be a sequence, then $\{x_n\}$ is called a **Cauchy sequence** provided that for each $\epsilon > 0$, there exists an integer $N > 0$ so that $m, n > N$ implies $\rho(x_n, x_m) < \epsilon$.*

5.16. Definition. *Let X be a metric space, then X is **complete** provided every Cauchy sequence in X converges to a point x_0 of X.*

EXAMPLES (AND EXERCISES)

5.26. The real line is a complete metric space with metric

$$\rho(x, y) = |x - y|.$$

5.27. Euclidean n-space, E^n, the topological product of n spaces each of which is the real line, is a complete metric space with metric

$$\rho(x, y) = \left[\sum_{i=1}^{n} |x_i - y_i|^2\right]^{1/2}.$$

5.28. Hilbert space is a complete metric space.

5.29. Let

$$C[0, 1] = \{f \mid f \text{ a continuous real-valued function with domain } [0, 1]\},$$

and let

$$\rho(f, g) = \sup_{0 \leq x \leq 1} |f(x) - g(x)|,$$

then $C[0, 1]$ is a complete metric space.

5.30. Let m be the set of all real-valued bounded sequences, i.e.,

$$m = \{x = \{x_n\} \mid x_n \text{ real for each } n = 1, 2, \ldots, |x_n| < K_x \text{ for some real } K_x \text{ depending upon } x\}.$$

Define

$$\rho(x, y) = \sup_n |x_n - y_n|,$$

then m is a complete metric space.

5.31. Let c be the set of all real convergent sequences; define $\rho(x, y) = \sup_n |x_n - y_n|$, then c is a complete metric space, and is furthermore a subspace of the space m of the preceding exercise.

5.32. Show that the definition of a complete metric space may be weakened by insisting only that every Cauchy sequence have a limit point in X, i.e., show that if a Cauchy sequence has a limit point x_0 in X, then it converges to x_0.

5.33. Show that every compact space is complete. [*Hint.* Use Ex. 5.32 above and 3.26.]

5.34. Show that any closed subspace of a complete metric space is complete.

5.35. (a) Let R be the reals, define $f : R \to (-1, 1)$ by

$$f(x) = \frac{x}{1 + |x|},$$

then f is a homeomorphism.

(b) R is a complete metric space, but the subspace $(-1, 1)$ to which R is homeomorphic [by part (a)] is not complete, since the Cauchy sequence

$$\left\{ \frac{n}{1 + n} \mid n = 1, 2, \ldots \right\}$$

does not converge to a point of $(-1, 1)$.

5.36. Completeness is not a topological property. [*Hint.* Use Ex. 5.35 above.]

5.37. (a) Let X be a metric space, and let $\{x_n\}$ be a sequence in X such that $\lim_n x_n = x_0 \in X$. Show that $\{x_n\}$ is a Cauchy sequence.

(b) Show that if $Y \subseteq X$, Y a subspace of X, then if Y is complete, Y is closed in X.

(c) In particular if $X = R$, then any complete bounded subset of X is compact. [A set $A \subseteq X$ is said to be bounded if there exists a point $x \in X$ and a finite real number N such that $A \subseteq S_N(x)$.]

5.38. Let $A = \{x \mid x_i = \delta_j^i, j = 1, 2, \ldots\} \subseteq \mathcal{K}$ be the set of unit points of Hilbert space (Cf. Ex. 5.22). Show that A is complete and bounded but not compact.

The situation encountered in Ex. 5.37 above is somewhat more general, namely in any Euclidean space $E^n = \underset{i=1}{\overset{n}{\times}} R_i$, where each R_i is the reals, a complete bounded subspace is compact. Unfortunately, this result does not hold in general, as Ex. 5.38 above shows us. We need not be dismayed, however, for by a slight strengthening of hypotheses we may overcome this difficulty.

We need first

5.17. Definition. *A metric space is **totally bounded** if for each $\epsilon > 0$, there exists a finite set $F = \{x_1, x_2, \ldots, x_{n_\epsilon}\}$ such that $X \subseteq \bigcup_{i=1}^{n_\epsilon} S_\epsilon(x_i)$.*

Then we have

5.18. Theorem. *Every complete totally bounded metric space is compact and conversely.*

Proof. Let X be complete and totally bounded. We exploit 5.10 and prove merely that X is countably compact. To that end, let A be an infinite subset of X. Since $\frac{1}{2} > 0$, there exists

$$F_1 = \{x_{11}, x_{12}, \ldots, x_{1n_1}\}$$

such that $X \subseteq \bigcup_{i=1}^{n_1} S_{1/2}(x_{1i})$ by the total boundedness of X. For some i, $S_{1/2}(x_{1i}) \cap A$ is an infinite set, otherwise A would be finite. We can assume $S_{1/2}(x_{11}) \cap A$ is infinite by suitably renaming the points of F_1. Let $A_1 = A \cap S_{1/2}(x_{11})$.

Now there exists $F_2 = \{x_{21}, x_{22}, \ldots, x_{2n_2}\}$ such that

$$X \subseteq \bigcup_{i=1}^{n_2} S_{1/4}(x_{2i}).$$

Since A_1 is infinite there exists an i such that $S_{1/4}(x_{2i}) \cap A_1$ is infinite, and again we may assume $A_2 = S_{1/4}(x_{21}) \cap A_1$ is the infinite set. Thus in general having defined A_{k-1}, there exists $F_k = \{x_{k1}, x_{k2}, \ldots, x_{kn_k}\}$ such that

$$X \subseteq \bigcup_{i=1}^{n_k} S_{1/2^k}(x_{ki}),$$

and $A_{k-1} \cap S_{1/2^k}(x_{ki})$ is infinite for some choice of i, which we assume again to be 1. We let $A_k = A_{k-1} \cap S_{1/2^k}(x_{k1})$. We note that, in general, $A_k \subseteq A_{k-1}$ for each $k > 1$.

Now select $x_1 \in A_1$, $x_2 \in A_2 - \{x_1\}$, and in general

$$x_k \in A_k - \bigcup_{i=1}^{k-1} \{x_i\}.$$

Since each A_k is infinite, this selection will always be possible. The sequence $\{x_n\}$ thus selected is a Cauchy sequence, for let $\epsilon > 0$, and select $k > 0$ so that $1/2^{k-1} < \epsilon$. Then if $m, n \geq k$,

$$x_m \in A_m \subseteq A_k \quad \text{and} \quad x_n \in A_n \subseteq A_k,$$

while $A_k \subseteq S_{1/2^k}(x_{k1})$. Consequently $x_n, x_m \in S_{1/2^k}(x_{k1})$ and

$$\rho(x_n, x_m) \leq \rho(x_n, x_{k1}) + \rho(x_{k1}, x_m)$$
$$< \frac{1}{2^k} + \frac{1}{2^k} = \frac{1}{2^{k-1}} < \epsilon.$$

Since X is complete, $\{x_n\}$ converges to some point $x_0 \in X$, and since $\{x_n\} \subseteq A$, x_0 is a limit point of A. Thus X is countably compact, and as remarked earlier 5.10 completes the argument.

The converse is relatively trivial, and we leave its proof to the student. ∎

§ 5 Category Theorems

We now investigate a sequence of theorems which lead to the so-called Baire Category theorem (sometimes called the Baire-Moore theorem), which finds occasional use in both topology and analysis.

We need first some definitions.

5.19. Definition. *Let $N \subseteq X$, a topological space, then N is called* **nowhere dense** *if $(\bar{N})^\circ = \varnothing$ (i.e., if the interior of the closure of N is empty).*

EXERCISE

5.39. If $N \subseteq X$ is nowhere dense then
  (a) $(\bar{N})^c$ is dense in X.
  (b) For each $x \in X$ and each open set O such that $x \in O$, there exists O_1, open and nonempty, such that $O_1 \subseteq O$ and $O_1 \subseteq N^c$.

5.20. Definition. *A set $B \subseteq X$, a topological space, is said to be of* **Baire Category** I *(or of the first category) if B is the union of countably many nowhere dense sets. A set is said to be of* **Baire Category** II *(or of the* second category) if it is not of Baire Category I. A set is said to be **residual** if its complement in X is of Baire Category I.

5.21. Theorem. *Let X be a metric space, then X is complete if and only if for each sequence $\{C_n\}$ of closed sets such that $C_{n+1} \subseteq C_n$ for each n, and such that*
$$\lim_n \delta(C_n) = 0,$$
$\bigcap_{n=1}^{\infty} C_n$ *is a single point.*

NOTE: $\delta(A) = \sup_{x,y \in A} \rho(x, y)$. (Cf. Ex. 5.6.)

Proof. Suppose X is complete. Construct a sequence
$$\{x_n \mid x_n \in C_n, n = 1, 2, \ldots\},$$
then $\{x_n\}$ is a Cauchy sequence, for let $\epsilon > 0$. Select $N_\epsilon > 0$ so that $n > N_\epsilon$ implies $\delta(C_n) < \epsilon$. Then for $n \geq m > N_\epsilon$, $x_n \in C_n \subseteq C_m$, and $x_m \in C_m$, whence
$$\rho(x_n, x_m) \leq \delta(C_m) < \epsilon.$$
Since X is complete, $\{x_n\}$ converges to some point $x_0 \in X$.

Now if $x_n = x_0$ for all $n > N$, for some $N > 0$, then
$$x_0 = x_n \in C_n \subseteq C_k$$
for $1 \leq k < n$ and $x_0 \in \bigcap_{n=1}^{\infty} C_n$.

On the other hand, if there exist arbitrarily large values of n such that $x_n \neq x_0$, then let $m > 0$ be fixed. Then for $\epsilon > 0$, there is an $n > m$, such that
$$x_n \neq x_0, \quad \rho(x_n, x_0) < \epsilon,$$
since $\lim_n x_n = x_0$. Thus x_0 is a limit point of C_n, and since C_n is closed, $x_0 \in C_n$. Also since $C_n \subseteq C_m$, $x_0 \in C_m$, and since m was arbitrary, $x_0 \in C_m$ for all m, consequently
$$x_0 \in \bigcap_{n=1}^{\infty} C_n.$$

Suppose now that $y \in \bigcap_{n=1}^{\infty} C_n$. Let $\epsilon > 0$, and select N_ϵ so that $n > N_\epsilon$ implies $\delta(C_n) < \epsilon$, then
$$x_0, y \in C_n \quad \text{and} \quad \rho(x_0, y) \leq \delta(C_n) < \epsilon.$$
Since ϵ was arbitrary, $\rho(x_0, y) = 0$, and $x_0 = y$. Thus
$$\bigcap_{n=1}^{\infty} C_n = \{x_0\}.$$

Conversely suppose that the condition of the theorem holds. Let $\{x_n\}$ be a Cauchy sequence in X, and let $S_\epsilon(x) = \{y \mid \rho(x, y) < \epsilon\}$. For each positive integer k, there exists a positive integer n_k such that $n \geq n_k$ implies $\rho(x_{n_k}, x_n) < 1/2^k$, and we may assume n_k so chosen that it is the least integer with this property. It then follows that $n_k \leq n_{k+1}$. Define
$$C_k = \overline{S_{1/2^{k-1}}(x_{n_k})}, \quad \text{for } k = 1, 2, \ldots.$$
Observe that
$$C_k \subseteq \left\{ y \mid \rho(x_{n_k}, y) \leq \frac{1}{2^{k-1}} \right\}$$
so that $\delta(C_k) \leq 1/2^{k-2}$ and $\lim_k \delta(C_k) = 0$.

We show that C_k is a contracting sequence, i.e., $C_{k+1} \subseteq C_k$. Let $y \in C_{k+1}$, then
$$\rho(y, x_{n_{k+1}}) \leq \frac{1}{2^k} \quad \text{and} \quad \rho(x_{n_k}, x_{n_{k+1}}) \leq \frac{1}{2^k}$$
by choice of n_k. Thus
$$\rho(y, x_{n_k}) \leq \rho(y, x_{n_{k+1}}) + \rho(x_{n_{k+1}}, x_{n_k}) \leq \frac{1}{2^k} + \frac{1}{2^k} = \frac{1}{2^{k-1}},$$
and $y \in C_k$. Therefore $C_{k+1} \subseteq C_k$.

By hypothesis $\bigcap_{k=1}^{\infty} C_k = \{x_0\}$ is a single point. Let $\epsilon > 0$, and select k so that $1/2^{k-2} < \epsilon$, then for $n > n_k$, since $x_0 \in C_k$, $\rho(x_0, x_n) \leq 1/2^{k-1}$, and

$$\rho(x_0, x_n) \leq \rho(x_0, x_{n_k}) + \rho(x_{n_k}, x_n) \leq \frac{1}{2^{k-1}} + \frac{1}{2^k} < \frac{1}{2^{k-2}} < \epsilon.$$

Thus $\lim_n x_n = x_0$, and X is complete. ∎

5.22. Theorem. *Let X be a complete metric space, then X is of Baire Category* II.

Proof. Let $N \subseteq X$, N of Category I. We need only show that $N \neq X$, or that there is a point $x \in X - N$.

$N = \bigcup_{n=1}^{\infty} N_n$, where N_n is nowhere dense. Now $(\overline{N_n})^c$ is dense in X, for each n, thus there is a point $x_1 \in (\overline{N_1})^c$ and there is an $\epsilon > 0$ so that $S_\epsilon(x_1) \cap N_1 = \emptyset$ by Ex. 5.39. Thus

$$\overline{S_{\epsilon/2}(x_1)} \subseteq S_\epsilon(x_1) \quad \text{and} \quad \overline{S_{\epsilon/2}(x_1)} \cap N_1 = \emptyset.$$

Let $\epsilon/2 = \epsilon_1$, then again since $(\overline{N_2})^c$ is dense in X, there exists a point $x_2 \in S_{\epsilon_1}(x_1) \cap (\overline{N_2})^c$. As before we select $\epsilon_2 > 0$ so that
(1) $0 < \epsilon_2 < \epsilon_1/2$,
(2) $\overline{S_{\epsilon_2}(x_2)} \cap N_2 = \emptyset$,
(3) $\overline{S_{\epsilon_2}(x_2)} \subseteq \overline{S_{\epsilon_1}(x_1)}$,
and in general for each n, select $x_n \in S_{\epsilon_{n-1}}(x_{n-1}) \cap N_n$, and ϵ_n so that
(1) $0 < \epsilon_n < \epsilon_{n-1}/2$,
(2) $\overline{S_{\epsilon_n}(x_n)} \cap N_n = \emptyset$,
(3) $\overline{S_{\epsilon_n}(x_n)} \subseteq \overline{S_{\epsilon_{n-1}}(x_{n-1})}$.

We thus obtain a sequence $\overline{S_{\epsilon_n}(x_n)}$ of contracting closed sets such that

$\delta\, \overline{(S_{\epsilon_n}(x_n))} < \epsilon/2^n \to 0$, whence by 5.21 there is a point $x_0 \in \bigcap_{n=1}^{\infty} \overline{S_{\epsilon_n}(x_n)}$. Let $m > 0$ be fixed, then

$$x_0 \in \overline{S_{\epsilon_m}(x_m)} \quad \text{and} \quad \overline{S_{\epsilon_m}(x_m)} \cap N_m = \emptyset,$$

thus $x_0 \notin N_m$, and since m was arbitrary, $x_0 \notin N_m$ for any m. Thus

$$x_0 \notin \bigcup_{n=1}^{\infty} N_n = N, \quad x_0 \in X - N,$$

whence X is not of Category I. ∎

5.23. Theorem. *Let X be a complete metric space, $R \subseteq X$, R residual, then R is dense.*

Proof. Let $x \in X - R$, we need only show that x is a limit point of R. Now $X - R$ is of Category I, thus

$$X - R = \bigcup_{n=1}^{\infty} N_n,$$

where each N_n is nowhere dense. Let O be an open set such that $x \in O$, and select $\epsilon > 0$ so that $S_\epsilon(x) \subseteq O$. We now proceed to construct a sequence $\{x_n\}$ of points and closures of neighborhoods of these points, $S_{\epsilon_n}(x_n)$ as in the proof of 5.22, so that

(1) $0 < \epsilon_n < \epsilon_{n-1}/2$,
(2) $\overline{S_{\epsilon_n}(x_n)} \cap N_n = \varnothing$,
(3) $\overline{S_{\epsilon_n}(x_n)} \subseteq \overline{S_{\epsilon_{n-1}}(x_{n-1})}$,

and further so that $x_1 \in S_\epsilon(x)$, and $\overline{S_{\epsilon_1}(x_1)} \subseteq S_\epsilon(x)$. This is possible, since x_1 was chosen in $(\overline{N_1})^c$, a dense subset of X, thus there is an $x_1 \in S_\epsilon(x) \cap (\overline{N_1})^c$, and we may choose ϵ_1 so that $\overline{S_{\epsilon_1}(x_1)} \subseteq S_\epsilon(x)$. Then as in 5.22

$$x_0 \in \bigcap_{n=1}^{\infty} \overline{S_{\epsilon_n}(x_n)} \subseteq S_\epsilon(x),$$

and $x_0 \notin N$. Thus there is an $x_0 \in X - N = R$ in each open set O which contains x. Consequently $x \in \bar{R}$, and R is dense in X. ∎

EXAMPLES (AND EXERCISES)

5.40. In the space $C[0, 1]$ of Ex. 5.29 which we know to be complete, let

$$P_m = \{f \mid \text{for some } x \in [0, 1], \text{ for the positive integer } m,$$
$$\text{and for all } h > 0, \ |(f(x + h) - f(x))/h| \leq m\}.$$

Let $S = C[0, 1] - \bigcup_{m=1}^{\infty} P_m$, then S is the set of all continuous functions on $[0, 1]$ not having a finite right-hand derivative at any point of $[0, 1]$.

(a) Show that each P_m is closed.
(b) Show that for each $f \in P_m$ and each $\epsilon > 0$ there exists a

$g \in C[0, 1]$ such that $g \notin P_m$, $g \in S_\epsilon(f)$. [*Hint.* Approximate f to within $\epsilon/2$ by a polygonal line, then approximate each segment of the polygonal line (to within $\epsilon/2$) by a sawtooth with teeth having slopes, both positive and negative, greater in absolute value than m.]

(c) Use (a) and (b) above to show that each P_m is nowhere dense.

(d) Show that $\bigcup_{m=1}^{\infty} P_m$ is of Baire Category I.

(e) Show that $S \neq \emptyset$.

(f) Prove that there exist functions continuous on $[0, 1]$ which do not have a finite derivative at any point of $[0, 1]$.

5.41. (a) Let X be a complete metric space and let $f : X \to X$ be a mapping such that $\rho(f(x), f(y)) < \alpha\rho(x, y)$, where $0 < \alpha < 1$ and α does not depend upon x or y. Show that there exists a point $x_0 \in X$, such that $f(x_0) = x_0$. [*Hint.* Let $x \in X$, define $x_1 = x, x_2 = f(x_1), \ldots, x_n = f(x_{n-1}), \ldots$. Show that $\{x_n\}$ is a Cauchy sequence, and let $x_0 = \lim_n x_n$. Show x_0 has the property $f(x_0) = x_0$.]

(b) Show further that x_0 is unique, i.e., there is no other point x_1 such that $f(x_1) = x_1$.

(c) Consider the differential equation $y' = f(x, y)$, with boundary condition $y(x_0) = y_0$, where x and y are real. Let f be continuous in

$$D = \{(x, y) \mid |x - x_0| \leq a, |y - y_0| \leq b, a > 0, b > 0\},$$

let $N = \max_{x,y \in D} |f(x, y)|$ and let $aN < b$. Further let f satisfy the Lipschitz condition:

$$|f(x, y_1) - f(x, y_2)| \leq K|y_1 - y_2|$$

for some fixed $K > 0$ and for $(x, y_1), (x, y_2) \in D$.

Now define

$$X = \{y(x) \mid y \text{ a continuous real-valued function of the real variable } x, \text{ and } |y(x) - y_0| \leq b \text{ for } |x - x_0| \leq a\},$$

and define

$$\rho(y_1, y_2) = \sup_{|x-x_0| \leq a} |y_1(x) - y_2(x)| \quad \text{for } y_1, y_2 \in X.$$

Show that X is a complete metric space with the metric ρ.

(d) Let X be as in part (c) above. Define $T: X \to X$ by

$$T(y) = y_0 + \int_{x_0}^{x} f(z, y(z))\, dz.$$

Show that $T(y) \in X$. Show further that

$$\rho(T(y_1), T(y_2)) \leq aK\rho(y_1, y_2).$$

Now select a so small that $aK < 1$.

(e) [Continuation of (d)] Show that there is a unique $y \in X$ such that $T(y) = y$, i.e.,

$$y(x) = y_0 + \int_{x_0}^{x} f(z, y(z))\, dz,$$

and hence that the differential equation $y'(x) = f(x, y)$ with boundary condition $y(x_0) = y_0$ has a unique solution in at least that portion of D for which $a < 1/K$.

5.42. (*Term paper*) It is the purpose of this exercise to generalize the technique that is often used to obtain the reals from the rationals. Specifically we want to prove that any metric space (the rationals, for example) is isometric to some subspace of a complete metric space (the reals, for example).

Now let X be a metric space with metric ρ. Let

$$Y = \{\{x_n\} \mid \{x_n\} \text{ a Cauchy sequence of points } x_n \in X\}.$$

Define a relation, "$\sim$," in Y, by

$$x = \{x_n\} \sim y = \{y_n\}$$

if and only if $\lim_n \rho(x_n, y_n) = 0$.

(a) Show that "$\sim$" is an equivalence relation in Y.

(b) Let $Z = Y/\sim$, i.e., Z is the set of all equivalence classes of elements of Y under the equivalence relation, "$\sim$." Denote an element $\bar{x} \in Z$ by $\bar{x} = [x]$, where $x \in Y$, $x = \{x_n\}$, $\{x_n\}$ a Cauchy sequence in X. In Z define

$$\bar\rho(\bar x, \bar y) = \lim_n \rho(x_n, y_n).$$

Then
  (i) Show that $\bar\rho$ is independent of the choice of representatives $\{x_n\} \in [x]$ and $\{y_n\} \in [y]$.
  (ii) Show that $\bar\rho$ is a metric for X.
(c) Let Z be provided with the metric topology induced by $\bar\rho$. Show that Z is complete.
(d) Define $f : X \to Z$ by $f(x) = [x] = [\{x_n\}]$, where $x_n = x$ for all n, i.e., every point $x \in X$ is mapped into a constant sequence, $x, x, x, \ldots$. Show that f is an isometry of X with a subset of Z.
(e) Show that $f(X)$ is dense in Z.
(f) Show that X is separable if and only if Z is separable.

REFERENCES

The following list of books is intended as a guide to further reading in point set topology. Some of the books cover the same material as is covered here; however, a number of them go considerably beyond what is attempted here, either exploring point set topology further, or investigating the application of point set topology to algebraic topology or analysis. Although each of the books listed below has its merits, the author has found the book by Kelley and the one by Hocking and Young to be particularly useful.

1. Alexandroff, P., and H. Hopf, *Topologie* (Ann Arbor, Mich.: Edwards, 1945).

2. Arnold, B. H., *Intuitive Concepts in Elementary Topology* (Englewood Cliffs, N.J.: Prentice-Hall, 1962).

3. Bourbaki, N., *Topologie Générale* (Paris: Actualités Scientifiques et Industrielles, Herman et Cie., 858 (1940) = 1152 (1951), 916 (1942) = 1143 (1951), 1029 (1947), 1045 (1948), 1084 (1949)).

4. Hall, D. W., and G. L. Spencer, *Elementary Topology* (New York: Wiley, 1955).

5. Hausdorff, F., *Mengenlehre* (Berlin: de Gruyter, 1927, 1935).

6. Hocking, J. G., and G. S. Young, *Topology* (Reading, Mass.: Addison-Wesley, 1961).

7. Hurewicz, W., and H. Wallman, *Dimension Theory* (Princeton, N.J.: Princeton, 1941).

8. Kelley, J. L., *General Topology* (Princeton, N.J.: Van Nostrand, 1955).

9. Kuratowski, K., *Introduction to Set Theory and Topology* (New York: Pergamon, 1961).

10. Kuratowski, K., *Topologie*, vols. 1 and 2 (2nd ed.; Warsaw, 1948).

11. Mansfield, M. J., *Introduction to Topology* (Princeton, N.J.: Van Nostrand, 1963).

12. Moore, R. L., *Foundations of Point Set Theory* (New York: American Mathematical Society Colloquium Publication No. 13, 1932).

13. Newman, M. H. A., *Elements of the Topology of Plane Sets of Points* (New York: Cambridge U.P., 1939).

14. Patterson, E. M., *Topology* (New York: Interscience, 1959).

15. Sierpinski, W., *Introduction to General Topology* (Toronto: University of Toronto Press, 1934, 1952).

16. Simmons, G. F., *Introduction to Topology and Modern Analysis* (New York: McGraw, 1963).

17. Vaidynathaswamy, R., *Treatise on Set Topology, Part 1* (Madras: Indian Mathematical Society, 1947).

18. Whyburn, G. T., *Analytic Topology* (New York: American Mathematical Society Colloquium Publication No. 28, 1942).

19. Wilder, R. L., *Topology of Manifolds* (New York: American Mathematical Society Colloquium Publication No. 32, 1949).

INDEX

ARC, 109
Arcwise connected, 109, 110
Axiom:
  of choice, 12
  separation, 79

BAIRE category, 136
Baire-Moore Theorem, 136
Base (basis):
  countable, 47
  for a topology, 43
Bound:
  greatest lower, 17
  least upper, 17
  lower, 17
  upper, 17
Bounded:
  set, 134
  totally, 135

CANTOR set, 112
Cartesian product, 6, 65
Category, 136
Cauchy sequence, 132
Chain, 101
  simple, 101
Chained by, 101
Choice:
  axiom of, 12
  function, 12
Closed:
  function, 62
  set, 31
Closure of a set, 28
Compact, 70
Complement of a set, 4
Complete, 132
Completely normal space, 87
Component, 105, 106

Connected, 99
  arcwise, 109, 110
  locally, 107
Constant function, 56
Continuous, 58
  image, 60
  at a point, 58
Countable:
  base (basis), 47
  set, 11
Countability:
  first axiom of, 49
  second axiom of, 47
Countably:
  compact, 89
  infinite, 11
Cover, 69
Cut point, 113

DE MORGAN'S rules, 4, 6
Dense, 33
  in itself, 103, 113
Derived set, 28
Diameter, 118
Difference, 4
Discrete topology, 23
Distance:
  from one set to another, 119
  from a point to a set, 119

EMPTY set, 2
Equivalence:
  class, 9
  relation, 9
Euler–Venn diagram, 7, 8
Extension of a function, 56

FACTOR space, 64
Finer topology, 42

Finite intersection property, 71
First axiom of countability, 49
Frontier of a set, 33
Function, 56
  choice, 12
  closed, 62
  constant, 56
  continuous, 58
    at a point, 58
  extension of, 56
  identity, 56
  interior, 62
  one-to-one, 11, 57
  onto, 57
  open, 62
  restriction of, 56

GRAPH, 68
Greatest lower bound, 17

HAUSDORFF:
  metric, 124
  space, 40, 79
Hilbert:
  cube, 126
  space, 125
Homeomorphism, 61

IDENTITY function, 56
Image, 56
  continuous, 60
Inclusion, 2
  proper, 2
Infimum, 17
Infinite set, 11
Interior:
  function, 62
  of a set, 33
Intersection, 3, 5
Inverse image, 57
Isometry, 118

KRONECKER delta, 126

LATTICE, 54
Least upper bound, 17
Limit:
  inferior of a sequence of sets, 39
  point, 27
  of a sequence of points, 38

Limit (cont.):
  of a sequence of sets, 39
  superior of a sequence of sets, 39
Lindelöf Theorem, 87
Line, usual topology for, 22
Lipschitz condition, 141
Locally:
  compact, 92
  connected, 107
  finite, 96
Lower bound, 17

MAPPING, 60
Mathematical induction, 16
Metric, 115
  Hausdorff, 124
  set, 115
  space, 116
  topology, 116
Metrizable space, 117

NEIGHBORHOOD system, 20
Normal space, 81
Nowhere dense, 136

ONE-TO-ONE correspondence, 11
One-to-one function, 11, 57
One-point compactification, 94
Onto, 57
Open:
  function, 62
  set, 23
Order relation, 10

PARACOMPACT, 96
Partial order relation, 10
Partition, 9
Plane, usual topology for, 22
Product:
  Cartesian, 6, 65
  space, 52, 65–68
  topology, 52, 65–68
Projection, 64, 66
  stereographic, 92, 93
Property:
  finite intersection, 71
  topological, 62

REFINEMENT, 96
Reflexivity, 9

INDEX

Regular space, 81
Relation, 9, 10
  equivalence, 9
  order, 10
    partial, 10
    simple, 10
Relative topology, 36
Residual, 137
Restriction of a function, 56
Right order topology, 23

SECOND axiom of countability, 47
Separable, 47
Separated, 99
Separation axioms, 79
Sequence, Cauchy, 132
Set(s):
  bounded, 134
  Cantor, 112
  Cartesian product of, 6, 65
  closed, 31
  closure of, 28
  complement of, 4
  countable, 11
  countably infinite, 11
  dense, 33
  derived, 28
  difference of, 4
  elements of, 2
  equality of, 2
  finite, 11
  frontier of, 33
  inclusion, 2
  indexing, 5
  infinite, 11
  interior of, 33
  intersection of, 3, 5
  metric, 115
  notation for, 1
  open, 23
  proper inclusion, 2
  separated, 99
  union of, 3, 5
Simple:
  chain, 101
  order relation, 10
Simply chained by, 101
Space:
  arcwise connected, 109, 110
  compact, 70

Space (*cont.*):
  complete metric, 132
  completely normal, 87
  connected, 99–104
  countably compact, 89
  factor, 64
  Hausdorff, 40, 79
  Hilbert, 125
  locally compact, 92
  locally connected, 107
  metric, 116
  metrizable, 117
  normal, 81
  product, 52, 65–68
  regular, 81
  separable, 47
  T_0, 79
  T_1, 79
  T_2, 79
  T_3, 79
  T_4, 79
  topological, 21
Stereographic projection, 92, 93
Sub-base for a topology, 50
Subset, 2
Subspace, 36
Supremum, 17
Symmetry, 9
System of neighborhoods, 20

TOPOLOGICAL:
  property, 62
  space, 21
Topology, 21
  base (basis) for, 43
  comparison of, 42
  discrete, 23
  product, 52, 65–68
  relative, 36
  right order, 23
  sub-base for, 50
  trivial, 23
  usual:
    for the plane, 22
    for the reals, 22
Totally bounded, 135
Totally disconnected, 111
Transitivity, 9
Trivial topology, 23

Tychonoff:
   plank, 84–87
   theorem, 77

UNION, 3, 5
Upper bound, 17

Urysohn's lemma, 127

WELL-ORDERED, 12

ZORN's lemma, 75

A CATALOG OF SELECTED
DOVER BOOKS
IN SCIENCE AND MATHEMATICS

CATALOG OF DOVER BOOKS

Astronomy

BURNHAM'S CELESTIAL HANDBOOK, Robert Burnham, Jr. Thorough guide to the stars beyond our solar system. Exhaustive treatment. Alphabetical by constellation: Andromeda to Cetus in Vol. 1; Chamaeleon to Orion in Vol. 2; and Pavo to Vulpecula in Vol. 3. Hundreds of illustrations. Index in Vol. 3. 2,000pp. $6^{1}/_{8}$ x $9^{1}/_{4}$.
Vol. I: 0-486-23567-X
Vol. II: 0-486-23568-8
Vol. III: 0-486-23673-0

EXPLORING THE MOON THROUGH BINOCULARS AND SMALL TELESCOPES, Ernest H. Cherrington, Jr. Informative, profusely illustrated guide to locating and identifying craters, rills, seas, mountains, other lunar features. Newly revised and updated with special section of new photos. Over 100 photos and diagrams. 240pp. $8^{1}/_{4}$ x 11. 0-486-24491-1

THE EXTRATERRESTRIAL LIFE DEBATE, 1750–1900, Michael J. Crowe. First detailed, scholarly study in English of the many ideas that developed from 1750 to 1900 regarding the existence of intelligent extraterrestrial life. Examines ideas of Kant, Herschel, Voltaire, Percival Lowell, many other scientists and thinkers. 16 illustrations. 704pp. $5^{3}/_{8}$ x $8^{1}/_{2}$. 0-486-40675-X

THEORIES OF THE WORLD FROM ANTIQUITY TO THE COPERNICAN REVOLUTION, Michael J. Crowe. Newly revised edition of an accessible, enlightening book re-creates the change from an earth-centered to a sun-centered conception of the solar system. 242pp. $5^{3}/_{8}$ x $8^{1}/_{2}$. 0-486-41444-2

ARISTARCHUS OF SAMOS: The Ancient Copernicus, Sir Thomas Heath. Heath's history of astronomy ranges from Homer and Hesiod to Aristarchus and includes quotes from numerous thinkers, compilers, and scholasticists from Thales and Anaximander through Pythagoras, Plato, Aristotle, and Heraclides. 34 figures. 448pp. $5^{3}/_{8}$ x $8^{1}/_{2}$.
0-486-43886-4

A COMPLETE MANUAL OF AMATEUR ASTRONOMY: TOOLS AND TECHNIQUES FOR ASTRONOMICAL OBSERVATIONS, P. Clay Sherrod with Thomas L. Koed. Concise, highly readable book discusses: selecting, setting up and maintaining a telescope; amateur studies of the sun; lunar topography and occultations; observations of Mars, Jupiter, Saturn, the minor planets and the stars; an introduction to photoelectric photometry; more. 1981 ed. 124 figures. 25 halftones. 37 tables. 335pp. $6^{1}/_{2}$ x $9^{1}/_{4}$. 0-486-42820-8

AMATEUR ASTRONOMER'S HANDBOOK, J. B. Sidgwick. Timeless, comprehensive coverage of telescopes, mirrors, lenses, mountings, telescope drives, micrometers, spectroscopes, more. 189 illustrations. 576pp. $5^{5}/_{8}$ x $8^{1}/_{4}$. (Available in U.S. only.)
0-486-24034-7

STAR LORE: Myths, Legends, and Facts, William Tyler Olcott. Captivating retellings of the origins and histories of ancient star groups include Pegasus, Ursa Major, Pleiades, signs of the zodiac, and other constellations. "Classic."—Sky & Telescope. 58 illustrations. 544pp. $5^{3}/_{8}$ x $8^{1}/_{2}$. 0-486-43581-4

CATALOG OF DOVER BOOKS

Chemistry

THE SCEPTICAL CHYMIST: THE CLASSIC 1661 TEXT, Robert Boyle. Boyle defines the term "element," asserting that all natural phenomena can be explained by the motion and organization of primary particles. 1911 ed. viii+232pp. $5^3/_8$ x $8^1/_2$.
0-486-42825-7

RADIOACTIVE SUBSTANCES, Marie Curie. Here is the celebrated scientist's doctoral thesis, the prelude to her receipt of the 1903 Nobel Prize. Curie discusses establishing atomic character of radioactivity found in compounds of uranium and thorium; extraction from pitchblende of polonium and radium; isolation of pure radium chloride; determination of atomic weight of radium; plus electric, photographic, luminous, heat, color effects of radioactivity. ii+94pp. $5^3/_8$ x $8^1/_2$. 0-486-42550-9

CHEMICAL MAGIC, Leonard A. Ford. Second Edition, Revised by E. Winston Grundmeier. Over 100 unusual stunts demonstrating cold fire, dust explosions, much more. Text explains scientific principles and stresses safety precautions. 128pp. $5^3/_8$ x $8^1/_2$. 0-486-67628-5

MOLECULAR THEORY OF CAPILLARITY, J. S. Rowlinson and B. Widom. History of surface phenomena offers critical and detailed examination and assessment of modern theories, focusing on statistical mechanics and application of results in mean-field approximation to model systems. 1989 edition. 352pp. $5^3/_8$ x $8^1/_2$. 0-486-42544-4

CHEMICAL AND CATALYTIC REACTION ENGINEERING, James J. Carberry. Designed to offer background for managing chemical reactions, this text examines behavior of chemical reactions and reactors; fluid-fluid and fluid-solid reaction systems; heterogeneous catalysis and catalytic kinetics; more. 1976 edition. 672pp. $6^1/_8$ x $9^1/_4$. 0-486-41736-0 $31.95

ELEMENTS OF CHEMISTRY, Antoine Lavoisier. Monumental classic by founder of modern chemistry in remarkable reprint of rare 1790 Kerr translation. A must for every student of chemistry or the history of science. 539pp. $5^3/_8$ x $8^1/_2$. 0-486-64624-6

MOLECULES AND RADIATION: An Introduction to Modern Molecular Spectroscopy. Second Edition, Jeffrey I. Steinfeld. This unified treatment introduces upper-level undergraduates and graduate students to the concepts and the methods of molecular spectroscopy and applications to quantum electronics, lasers, and related optical phenomena. 1985 edition. 512pp. $5^3/_8$ x $8^1/_2$. 0-486-44152-0

A SHORT HISTORY OF CHEMISTRY, J. R. Partington. Classic exposition explores origins of chemistry, alchemy, early medical chemistry, nature of atmosphere, theory of valency, laws and structure of atomic theory, much more. 428pp. $5^3/_8$ x $8^1/_2$. (Available in U.S. only.) 0-486-65977-1

GENERAL CHEMISTRY, Linus Pauling. Revised 3rd edition of classic first-year text by Nobel laureate. Atomic and molecular structure, quantum mechanics, statistical mechanics, thermodynamics correlated with descriptive chemistry. Problems. 992pp. $5^3/_8$ x $8^1/_2$.
0-486-65622-5

ELECTRON CORRELATION IN MOLECULES, S. Wilson. This text addresses one of theoretical chemistry's central problems. Topics include molecular electronic structure, independent electron models, electron correlation, the linked diagram theorem, and related topics. 1984 edition. 304pp. $5^3/_8$ x $8^1/_2$. 0-486-45879-2

CATALOG OF DOVER BOOKS

Engineering

DE RE METALLICA, Georgius Agricola. The famous Hoover translation of greatest treatise on technological chemistry, engineering, geology, mining of early modern times (1556). All 289 original woodcuts. 638pp. 6¾ x 11. 0-486-60006-8

FUNDAMENTALS OF ASTRODYNAMICS, Roger Bate et al. Modern approach developed by U.S. Air Force Academy. Designed as a first course. Problems, exercises. Numerous illustrations. 455pp. 5⅜ x 8½. 0-486-60061-0

DYNAMICS OF FLUIDS IN POROUS MEDIA, Jacob Bear. For advanced students of ground water hydrology, soil mechanics and physics, drainage and irrigation engineering and more. 335 illustrations. Exercises, with answers. 784pp. 6⅛ x 9¼. 0-486-65675-6

THEORY OF VISCOELASTICITY (SECOND EDITION), Richard M. Christensen. Complete consistent description of the linear theory of the viscoelastic behavior of materials. Problem-solving techniques discussed. 1982 edition. 29 figures. xiv+364pp. 6⅛ x 9¼.
0-486-42880-X

MECHANICS, J. P. Den Hartog. A classic introductory text or refresher. Hundreds of applications and design problems illuminate fundamentals of trusses, loaded beams and cables, etc. 334 answered problems. 462pp. 5⅜ x 8½. 0-486-60754-2

MECHANICAL VIBRATIONS, J. P. Den Hartog. Classic textbook offers lucid explanations and illustrative models, applying theories of vibrations to a variety of practical industrial engineering problems. Numerous figures. 233 problems, solutions. Appendix. Index. Preface. 436pp. 5⅜ x 8½. 0-486-64785-4

STRENGTH OF MATERIALS, J. P. Den Hartog. Full, clear treatment of basic material (tension, torsion, bending, etc.) plus advanced material on engineering methods, applications. 350 answered problems. 323pp. 5⅜ x 8½. 0-486-60755-0

A HISTORY OF MECHANICS, René Dugas. Monumental study of mechanical principles from antiquity to quantum mechanics. Contributions of ancient Greeks, Galileo, Leonardo, Kepler, Lagrange, many others. 671pp. 5⅜ x 8½. 0-486-65632-2

STABILITY THEORY AND ITS APPLICATIONS TO STRUCTURAL MECHANICS, Clive L. Dym. Self-contained text focuses on Koiter postbuckling analyses, with mathematical notions of stability of motion. Basing minimum energy principles for static stability upon dynamic concepts of stability of motion, it develops asymptotic buckling and postbuckling analyses from potential energy considerations, with applications to columns, plates, and arches. 1974 ed. 208pp. 5⅜ x 8½. 0-486-42541-X

BASIC ELECTRICITY, U.S. Bureau of Naval Personnel. Originally a training course; best nontechnical coverage. Topics include batteries, circuits, conductors, AC and DC, inductance and capacitance, generators, motors, transformers, amplifiers, etc. Many questions with answers. 349 illustrations. 1969 edition. 448pp. 6½ x 9¼. 0-486-20973-3

CATALOG OF DOVER BOOKS

ROCKETS, Robert Goddard. Two of the most significant publications in the history of rocketry and jet propulsion: "A Method of Reaching Extreme Altitudes" (1919) and "Liquid Propellant Rocket Development" (1936). 128pp. 5⅜ x 8½. 0-486-42537-1

STATISTICAL MECHANICS: PRINCIPLES AND APPLICATIONS, Terrell L. Hill. Standard text covers fundamentals of statistical mechanics, applications to fluctuation theory, imperfect gases, distribution functions, more. 448pp. 5⅜ x 8½. 0-486-65390-0

ENGINEERING AND TECHNOLOGY 1650–1750: ILLUSTRATIONS AND TEXTS FROM ORIGINAL SOURCES, Martin Jensen. Highly readable text with more than 200 contemporary drawings and detailed engravings of engineering projects dealing with surveying, leveling, materials, hand tools, lifting equipment, transport and erection, piling, bailing, water supply, hydraulic engineering, and more. Among the specific projects outlined-transporting a 50-ton stone to the Louvre, erecting an obelisk, building timber locks, and dredging canals. 207pp. 8⅜ x 11¼. 0-486-42232-1

THE VARIATIONAL PRINCIPLES OF MECHANICS, Cornelius Lanczos. Graduate level coverage of calculus of variations, equations of motion, relativistic mechanics, more. First inexpensive paperbound edition of classic treatise. Index. Bibliography. 418pp. 5⅜ x 8½. 0-486-65067-7

PROTECTION OF ELECTRONIC CIRCUITS FROM OVERVOLTAGES, Ronald B. Standler. Five-part treatment presents practical rules and strategies for circuits designed to protect electronic systems from damage by transient overvoltages. 1989 ed. xxiv+434pp. 6⅛ x 9¼. 0-486-42552-5

ROTARY WING AERODYNAMICS, W. Z. Stepniewski. Clear, concise text covers aerodynamic phenomena of the rotor and offers guidelines for helicopter performance evaluation. Originally prepared for NASA. 537 figures. 640pp. 6⅛ x 9¼. 0-486-64647-5

INTRODUCTION TO SPACE DYNAMICS, William Tyrrell Thomson. Comprehensive, classic introduction to space-flight engineering for advanced undergraduate and graduate students. Includes vector algebra, kinematics, transformation of coordinates. Bibliography. Index. 352pp. 5⅜ x 8½. 0-486-65113-4

HISTORY OF STRENGTH OF MATERIALS, Stephen P. Timoshenko. Excellent historical survey of the strength of materials with many references to the theories of elasticity and structure. 245 figures. 452pp. 5⅜ x 8½. 0-486-61187-6

ANALYTICAL FRACTURE MECHANICS, David J. Unger. Self-contained text supplements standard fracture mechanics texts by focusing on analytical methods for determining crack-tip stress and strain fields. 336pp. 6⅛ x 9¼. 0-486-41737-9

STATISTICAL MECHANICS OF ELASTICITY, J. H. Weiner. Advanced, self-contained treatment illustrates general principles and elastic behavior of solids. Part 1, based on classical mechanics, studies thermoelastic behavior of crystalline and polymeric solids. Part 2, based on quantum mechanics, focuses on interatomic force laws, behavior of solids, and thermally activated processes. For students of physics and chemistry and for polymer physicists. 1983 ed. 96 figures. 496pp. 5⅜ x 8½. 0-486-42260-7

CATALOG OF DOVER BOOKS

Mathematics

FUNCTIONAL ANALYSIS (Second Corrected Edition), George Bachman and Lawrence Narici. Excellent treatment of subject geared toward students with background in linear algebra, advanced calculus, physics and engineering. Text covers introduction to inner-product spaces, normed, metric spaces, and topological spaces; complete orthonormal sets, the Hahn-Banach Theorem and its consequences, and many other related subjects. 1966 ed. 544pp. $6\frac{1}{8}$ x $9\frac{1}{4}$. 0-486-40251-7

DIFFERENTIAL MANIFOLDS, Antoni A. Kosinski. Introductory text for advanced undergraduates and graduate students presents systematic study of the topological structure of smooth manifolds, starting with elements of theory and concluding with method of surgery. 1993 edition. 288pp. $5\frac{3}{8}$ x $8\frac{1}{2}$. 0-486-46244-7

VECTOR AND TENSOR ANALYSIS WITH APPLICATIONS, A. I. Borisenko and I. E. Tarapov. Concise introduction. Worked-out problems, solutions, exercises. 257pp. $5\frac{3}{8}$ x $8\frac{1}{4}$. 0-486-63833-2

AN INTRODUCTION TO ORDINARY DIFFERENTIAL EQUATIONS, Earl A. Coddington. A thorough and systematic first course in elementary differential equations for undergraduates in mathematics and science, with many exercises and problems (with answers). Index. 304pp. $5\frac{3}{8}$ x $8\frac{1}{2}$. 0-486-65942-9

FOURIER SERIES AND ORTHOGONAL FUNCTIONS, Harry F. Davis. An incisive text combining theory and practical example to introduce Fourier series, orthogonal functions and applications of the Fourier method to boundary-value problems. 570 exercises. Answers and notes. 416pp. $5\frac{3}{8}$ x $8\frac{1}{2}$. 0-486-65973-9

COMPUTABILITY AND UNSOLVABILITY, Martin Davis. Classic graduate-level introduction to theory of computability, usually referred to as theory of recurrent functions. New preface and appendix. 288pp. $5\frac{3}{8}$ x $8\frac{1}{2}$. 0-486-61471-9

AN INTRODUCTION TO MATHEMATICAL ANALYSIS, Robert A. Rankin. Dealing chiefly with functions of a single real variable, this text by a distinguished educator introduces limits, continuity, differentiability, integration, convergence of infinite series, double series, and infinite products. 1963 edition. 624pp. $5\frac{3}{8}$ x $8\frac{1}{2}$. 0-486-46251-X

METHODS OF NUMERICAL INTEGRATION (SECOND EDITION), Philip J. Davis and Philip Rabinowitz. Requiring only a background in calculus, this text covers approximate integration over finite and infinite intervals, error analysis, approximate integration in two or more dimensions, and automatic integration. 1984 edition. 624pp. $5\frac{3}{8}$ x $8\frac{1}{2}$. 0-486-45339-1

INTRODUCTION TO LINEAR ALGEBRA AND DIFFERENTIAL EQUATIONS, John W. Dettman. Excellent text covers complex numbers, determinants, orthonormal bases, Laplace transforms, much more. Exercises with solutions. Undergraduate level. 416pp. $5\frac{3}{8}$ x $8\frac{1}{2}$. 0-486-65191-6

RIEMANN'S ZETA FUNCTION, H. M. Edwards. Superb, high-level study of landmark 1859 publication entitled "On the Number of Primes Less Than a Given Magnitude" traces developments in mathematical theory that it inspired. xiv+315pp. $5\frac{3}{8}$ x $8\frac{1}{2}$. 0-486-41740-9

CATALOG OF DOVER BOOKS

CALCULUS OF VARIATIONS WITH APPLICATIONS, George M. Ewing. Applications-oriented introduction to variational theory develops insight and promotes understanding of specialized books, research papers. Suitable for advanced undergraduate/graduate students as primary, supplementary text. 352pp. 5 3/8 x 8 1/2.
0-486-64856-7

MATHEMATICIAN'S DELIGHT, W. W. Sawyer. "Recommended with confidence" by *The Times Literary Supplement*, this lively survey was written by a renowned teacher. It starts with arithmetic and algebra, gradually proceeding to trigonometry and calculus. 1943 edition. 240pp. 5 3/8 x 8 1/2.
0-486-46240-4

ADVANCED EUCLIDEAN GEOMETRY, Roger A. Johnson. This classic text explores the geometry of the triangle and the circle, concentrating on extensions of Euclidean theory, and examining in detail many relatively recent theorems. 1929 edition. 336pp. 5 3/8 x 8 1/2.
0-486-46237-4

COUNTEREXAMPLES IN ANALYSIS, Bernard R. Gelbaum and John M. H. Olmsted. These counterexamples deal mostly with the part of analysis known as "real variables." The first half covers the real number system, and the second half encompasses higher dimensions. 1962 edition. xxiv+198pp. 5 3/8 x 8 1/2.
0-486-42875-3

CATASTROPHE THEORY FOR SCIENTISTS AND ENGINEERS, Robert Gilmore. Advanced-level treatment describes mathematics of theory grounded in the work of Poincaré, R. Thom, other mathematicians. Also important applications to problems in mathematics, physics, chemistry and engineering. 1981 edition. References. 28 tables. 397 black-and-white illustrations. xvii + 666pp. 6 1/8 x 9 1/4.
0-486-67539-4

COMPLEX VARIABLES: Second Edition, Robert B. Ash and W. P. Novinger. Suitable for advanced undergraduates and graduate students, this newly revised treatment covers Cauchy theorem and its applications, analytic functions, and the prime number theorem. Numerous problems and solutions. 2004 edition. 224pp. 6 1/2 x 9 1/4.
0-486-46250-1

NUMERICAL METHODS FOR SCIENTISTS AND ENGINEERS, Richard Hamming. Classic text stresses frequency approach in coverage of algorithms, polynomial approximation, Fourier approximation, exponential approximation, other topics. Revised and enlarged 2nd edition. 721pp. 5 3/8 x 8 1/2.
0-486-65241-6

INTRODUCTION TO NUMERICAL ANALYSIS (2nd Edition), F. B. Hildebrand. Classic, fundamental treatment covers computation, approximation, interpolation, numerical differentiation and integration, other topics. 150 new problems. 669pp. 5 3/8 x 8 1/2.
0-486-65363-3

MARKOV PROCESSES AND POTENTIAL THEORY, Robert M. Blumental and Ronald K. Getoor. This graduate-level text explores the relationship between Markov processes and potential theory in terms of excessive functions, multiplicative functionals and subprocesses, additive functionals and their potentials, and dual processes. 1968 edition. 320pp. 5 3/8 x 8 1/2.
0-486-46263-3

ABSTRACT SETS AND FINITE ORDINALS: An Introduction to the Study of Set Theory, G. B. Keene. This text unites logical and philosophical aspects of set theory in a manner intelligible to mathematicians without training in formal logic and to logicians without a mathematical background. 1961 edition. 112pp. 5 3/8 x 8 1/2.
0-486-46249-8

CATALOG OF DOVER BOOKS

INTRODUCTORY REAL ANALYSIS, A.N. Kolmogorov, S. V. Fomin. Translated by Richard A. Silverman. Self-contained, evenly paced introduction to real and functional analysis. Some 350 problems. 403pp. 5⅜ x 8½. 0-486-61226-0

APPLIED ANALYSIS, Cornelius Lanczos. Classic work on analysis and design of finite processes for approximating solution of analytical problems. Algebraic equations, matrices, harmonic analysis, quadrature methods, much more. 559pp. 5⅜ x 8½. 0-486-65656-X

AN INTRODUCTION TO ALGEBRAIC STRUCTURES, Joseph Landin. Superb self-contained text covers "abstract algebra": sets and numbers, theory of groups, theory of rings, much more. Numerous well-chosen examples, exercises. 247pp. 5⅜ x 8½. 0-486-65940-2

QUALITATIVE THEORY OF DIFFERENTIAL EQUATIONS, V. V. Nemytskii and V.V. Stepanov. Classic graduate-level text by two prominent Soviet mathematicians covers classical differential equations as well as topological dynamics and ergodic theory. Bibliographies. 523pp. 5⅜ x 8½. 0-486-65954-2

THEORY OF MATRICES, Sam Perlis. Outstanding text covering rank, nonsingularity and inverses in connection with the development of canonical matrices under the relation of equivalence, and without the intervention of determinants. Includes exercises. 237pp. 5⅜ x 8½. 0-486-66810-X

INTRODUCTION TO ANALYSIS, Maxwell Rosenlicht. Unusually clear, accessible coverage of set theory, real number system, metric spaces, continuous functions, Riemann integration, multiple integrals, more. Wide range of problems. Undergraduate level. Bibliography. 254pp. 5⅜ x 8½. 0-486-65038-3

MODERN NONLINEAR EQUATIONS, Thomas L. Saaty. Emphasizes practical solution of problems; covers seven types of equations. ". . . a welcome contribution to the existing literature. . . ."—*Math Reviews.* 490pp. 5⅜ x 8½. 0-486-64232-1

MATRICES AND LINEAR ALGEBRA, Hans Schneider and George Phillip Barker. Basic textbook covers theory of matrices and its applications to systems of linear equations and related topics such as determinants, eigenvalues and differential equations. Numerous exercises. 432pp. 5⅜ x 8½. 0-486-66014-1

LINEAR ALGEBRA, Georgi E. Shilov. Determinants, linear spaces, matrix algebras, similar topics. For advanced undergraduates, graduates. Silverman translation. 387pp. 5⅜ x 8½. 0-486-63518-X

MATHEMATICAL METHODS OF GAME AND ECONOMIC THEORY: Revised Edition, Jean-Pierre Aubin. This text begins with optimization theory and convex analysis, followed by topics in game theory and mathematical economics, and concluding with an introduction to nonlinear analysis and control theory. 1982 edition. 656pp. 6⅛ x 9¼. 0-486-46265-X

SET THEORY AND LOGIC, Robert R. Stoll. Lucid introduction to unified theory of mathematical concepts. Set theory and logic seen as tools for conceptual understanding of real number system. 496pp. 5⅜ x 8¼. 0-486-63829-4

TENSOR CALCULUS, J.L. Synge and A. Schild. Widely used introductory text covers spaces and tensors, basic operations in Riemannian space, non-Riemannian spaces, etc. 324pp. $5^3/_8$ x $8^1/_4$. 0-486-63612-7

ORDINARY DIFFERENTIAL EQUATIONS, Morris Tenenbaum and Harry Pollard. Exhaustive survey of ordinary differential equations for undergraduates in mathematics, engineering, science. Thorough analysis of theorems. Diagrams. Bibliography. Index. 818pp. $5^3/_8$ x $8^1/_2$. 0-486-64940-7

INTEGRAL EQUATIONS, F. G. Tricomi. Authoritative, well-written treatment of extremely useful mathematical tool with wide applications. Volterra Equations, Fredholm Equations, much more. Advanced undergraduate to graduate level. Exercises. Bibliography. 238pp. $5^3/_8$ x $8^1/_2$. 0-486-64828-1

FOURIER SERIES, Georgi P. Tolstov. Translated by Richard A. Silverman. A valuable addition to the literature on the subject, moving clearly from subject to subject and theorem to theorem. 107 problems, answers. 336pp. $5^3/_8$ x $8^1/_2$. 0-486-63317-9

INTRODUCTION TO MATHEMATICAL THINKING, Friedrich Waismann. Examinations of arithmetic, geometry, and theory of integers; rational and natural numbers; complete induction; limit and point of accumulation; remarkable curves; complex and hypercomplex numbers, more. 1959 ed. 27 figures. xii+260pp. $5^3/_8$ x $8^1/_2$. 0-486-42804-8

THE RADON TRANSFORM AND SOME OF ITS APPLICATIONS, Stanley R. Deans. Of value to mathematicians, physicists, and engineers, this excellent introduction covers both theory and applications, including a rich array of examples and literature. Revised and updated by the author. 1993 edition. 304pp. $6^1/_8$ x $9^1/_4$. 0-486-46241-2

CALCULUS OF VARIATIONS, Robert Weinstock. Basic introduction covering isoperimetric problems, theory of elasticity, quantum mechanics, electrostatics, etc. Exercises throughout. 326pp. $5^3/_8$ x $8^1/_2$. 0-486-63069-2

THE CONTINUUM: A CRITICAL EXAMINATION OF THE FOUNDATION OF ANALYSIS, Hermann Weyl. Classic of 20th-century foundational research deals with the conceptual problem posed by the continuum. 156pp. $5^3/_8$ x $8^1/_2$. 0-486-67982-9

CHALLENGING MATHEMATICAL PROBLEMS WITH ELEMENTARY SOLUTIONS, A. M. Yaglom and I. M. Yaglom. Over 170 challenging problems on probability theory, combinatorial analysis, points and lines, topology, convex polygons, many other topics. Solutions. Total of 445pp. $5^3/_8$ x $8^1/_2$. Two-vol. set.
Vol. I: 0-486-65536-9 Vol. II: 0-486-65537-7

INTRODUCTION TO PARTIAL DIFFERENTIAL EQUATIONS WITH APPLICATIONS, E. C. Zachmanoglou and Dale W. Thoe. Essentials of partial differential equations applied to common problems in engineering and the physical sciences. Problems and answers. 416pp. $5^3/_8$ x $8^1/_2$. 0-486-65251-3

STOCHASTIC PROCESSES AND FILTERING THEORY, Andrew H. Jazwinski. This unified treatment presents material previously available only in journals, and in terms accessible to engineering students. Although theory is emphasized, it discusses numerous practical applications as well. 1970 edition. 400pp. $5^3/_8$ x $8^1/_2$. 0-486-46274-9

CATALOG OF DOVER BOOKS

Math—Decision Theory, Statistics, Probability

INTRODUCTION TO PROBABILITY, John E. Freund. Featured topics include permutations and factorials, probabilities and odds, frequency interpretation, mathematical expectation, decision-making, postulates of probability, rule of elimination, much more. Exercises with some solutions. Summary. 1973 edition. 247pp. 5⅜ x 8½.
0-486-67549-1

STATISTICAL AND INDUCTIVE PROBABILITIES, Hugues Leblanc. This treatment addresses a decades-old dispute among probability theorists, asserting that both statistical and inductive probabilities may be treated as sentence-theoretic measurements, and that the latter qualify as estimates of the former. 1962 edition. 160pp. 5⅜ x 8½.
0-486-44980-7

APPLIED MULTIVARIATE ANALYSIS: Using Bayesian and Frequentist Methods of Inference, Second Edition, S. James Press. This two-part treatment deals with foundations as well as models and applications. Topics include continuous multivariate distributions; regression and analysis of variance; factor analysis and latent structure analysis; and structuring multivariate populations. 1982 edition. 692pp. 5⅜ x 8½. 0-486-44236-5

LINEAR PROGRAMMING AND ECONOMIC ANALYSIS, Robert Dorfman, Paul A. Samuelson and Robert M. Solow. First comprehensive treatment of linear programming in standard economic analysis. Game theory, modern welfare economics, Leontief input-output, more. 525pp. 5⅜ x 8½.
0-486-65491-5

PROBABILITY: AN INTRODUCTION, Samuel Goldberg. Excellent basic text covers set theory, probability theory for finite sample spaces, binomial theorem, much more. 360 problems. Bibliographies. 322pp. 5⅜ x 8½.
0-486-65252-1

GAMES AND DECISIONS: INTRODUCTION AND CRITICAL SURVEY, R. Duncan Luce and Howard Raiffa. Superb nontechnical introduction to game theory, primarily applied to social sciences. Utility theory, zero-sum games, n-person games, decision-making, much more. Bibliography. 509pp. 5⅜ x 8½.
0-486-65943-7

INTRODUCTION TO THE THEORY OF GAMES, J. C. C. McKinsey. This comprehensive overview of the mathematical theory of games illustrates applications to situations involving conflicts of interest, including economic, social, political, and military contexts. Appropriate for advanced undergraduate and graduate courses; advanced calculus a prerequisite. 1952 ed. x+372pp. 5⅜ x 8½.
0-486-42811-7

FIFTY CHALLENGING PROBLEMS IN PROBABILITY WITH SOLUTIONS, Frederick Mosteller. Remarkable puzzlers, graded in difficulty, illustrate elementary and advanced aspects of probability. Detailed solutions. 88pp. 5⅜ x 8½. 0-486-65355-2

PROBABILITY THEORY: A CONCISE COURSE, Y. A. Rozanov. Highly readable, self-contained introduction covers combination of events, dependent events, Bernoulli trials, etc. 148pp. 5⅜ x 8¼.
0-486-63544-9

THE STATISTICAL ANALYSIS OF EXPERIMENTAL DATA, John Mandel. First half of book presents fundamental mathematical definitions, concepts and facts while remaining half deals with statistics primarily as an interpretive tool. Well-written text, numerous worked examples with step-by-step presentation. Includes 116 tables. 448pp. 5⅜ x 8½.
0-486-64666-1

CATALOG OF DOVER BOOKS

Math—Geometry and Topology

ELEMENTARY CONCEPTS OF TOPOLOGY, Paul Alexandroff. Elegant, intuitive approach to topology from set-theoretic topology to Betti groups; how concepts of topology are useful in math and physics. 25 figures. 57pp. $5^{3}/_{8}$ x $8^{1}/_{2}$. 0-486-60747-X

A LONG WAY FROM EUCLID, Constance Reid. Lively guide by a prominent historian focuses on the role of Euclid's Elements in subsequent mathematical developments. Elementary algebra and plane geometry are sole prerequisites. 80 drawings. 1963 edition. 304pp. $5^{3}/_{8}$ x $8^{1}/_{2}$. 0-486-43613-6

EXPERIMENTS IN TOPOLOGY, Stephen Barr. Classic, lively explanation of one of the byways of mathematics. Klein bottles, Moebius strips, projective planes, map coloring, problem of the Koenigsberg bridges, much more, described with clarity and wit. 43 figures. 210pp. $5^{3}/_{8}$ x $8^{1}/_{2}$. 0-486-25933-1

THE GEOMETRY OF RENÉ DESCARTES, René Descartes. The great work founded analytical geometry. Original French text, Descartes's own diagrams, together with definitive Smith-Latham translation. 244pp. $5^{3}/_{8}$ x $8^{1}/_{2}$. 0-486-60068-8

EUCLIDEAN GEOMETRY AND TRANSFORMATIONS, Clayton W. Dodge. This introduction to Euclidean geometry emphasizes transformations, particularly isometries and similarities. Suitable for undergraduate courses, it includes numerous examples, many with detailed answers. 1972 ed. viii+296pp. $6^{1}/_{8}$ x $9^{1}/_{4}$. 0-486-43476-1

EXCURSIONS IN GEOMETRY, C. Stanley Ogilvy. A straightedge, compass, and a little thought are all that's needed to discover the intellectual excitement of geometry. Harmonic division and Apollonian circles, inversive geometry, hexlet, Golden Section, more. 132 illustrations. 192pp. $5^{3}/_{8}$ x $8^{1}/_{2}$. 0-486-26530-7

THE THIRTEEN BOOKS OF EUCLID'S ELEMENTS, translated with introduction and commentary by Sir Thomas L. Heath. Definitive edition. Textual and linguistic notes, mathematical analysis. 2,500 years of critical commentary. Unabridged. 1,414pp. $5^{3}/_{8}$ x $8^{1}/_{2}$. Three-vol. set.
Vol. I: 0-486-60088-2   Vol. II: 0-486-60089-0   Vol. III: 0-486-60090-4

SPACE AND GEOMETRY: IN THE LIGHT OF PHYSIOLOGICAL, PSYCHOLOGICAL AND PHYSICAL INQUIRY, Ernst Mach. Three essays by an eminent philosopher and scientist explore the nature, origin, and development of our concepts of space, with a distinctness and precision suitable for undergraduate students and other readers. 1906 ed. vi+148pp. $5^{3}/_{8}$ x $8^{1}/_{2}$. 0-486-43909-7

GEOMETRY OF COMPLEX NUMBERS, Hans Schwerdtfeger. Illuminating, widely praised book on analytic geometry of circles, the Moebius transformation, and two-dimensional non-Euclidean geometries. 200pp. $5^{3}/_{8}$ x $8^{1}/_{4}$. 0-486-63830-8

DIFFERENTIAL GEOMETRY, Heinrich W. Guggenheimer. Local differential geometry as an application of advanced calculus and linear algebra. Curvature, transformation groups, surfaces, more. Exercises. 62 figures. 378pp. $5^{3}/_{8}$ x $8^{1}/_{2}$. 0-486-63433-7

History of Math

THE WORKS OF ARCHIMEDES, Archimedes (T. L. Heath, ed.). Topics include the famous problems of the ratio of the areas of a cylinder and an inscribed sphere; the measurement of a circle; the properties of conoids, spheroids, and spirals; and the quadrature of the parabola. Informative introduction. clxxxvi+326pp. 5⅜ x 8½. 0-486-42084-1

A SHORT ACCOUNT OF THE HISTORY OF MATHEMATICS, W. W. Rouse Ball. One of clearest, most authoritative surveys from the Egyptians and Phoenicians through 19th-century figures such as Grassman, Galois, Riemann. Fourth edition. 522pp. 5⅜ x 8½. 0-486-20630-0

THE HISTORY OF THE CALCULUS AND ITS CONCEPTUAL DEVELOPMENT, Carl B. Boyer. Origins in antiquity, medieval contributions, work of Newton, Leibniz, rigorous formulation. Treatment is verbal. 346pp. 5⅜ x 8½. 0-486-60509-4

THE HISTORICAL ROOTS OF ELEMENTARY MATHEMATICS, Lucas N. H. Bunt, Phillip S. Jones, and Jack D. Bedient. Fundamental underpinnings of modern arithmetic, algebra, geometry and number systems derived from ancient civilizations. 320pp. 5⅜ x 8½. 0-486-25563-8

THE HISTORY OF THE CALCULUS AND ITS CONCEPTUAL DEVELOPMENT, Carl B. Boyer. Fluent description of the development of both the integral and differential calculus—its early beginnings in antiquity, medieval contributions, and a consideration of Newton and Leibniz. 368pp. 5⅜ x 8½. 0-486-60509-4

GAMES, GODS & GAMBLING: A HISTORY OF PROBABILITY AND STATISTICAL IDEAS, F. N. David. Episodes from the lives of Galileo, Fermat, Pascal, and others illustrate this fascinating account of the roots of mathematics. Features thought-provoking references to classics, archaeology, biography, poetry. 1962 edition. 304pp. 5⅜ x 8½. (Available in U.S. only.) 0-486-40023-9

OF MEN AND NUMBERS: THE STORY OF THE GREAT MATHEMATICIANS, Jane Muir. Fascinating accounts of the lives and accomplishments of history's greatest mathematical minds—Pythagoras, Descartes, Euler, Pascal, Cantor, many more. Anecdotal, illuminating. 30 diagrams. Bibliography. 256pp. 5⅜ x 8½. 0-486-28973-7

HISTORY OF MATHEMATICS, David E. Smith. Nontechnical survey from ancient Greece and Orient to late 19th century; evolution of arithmetic, geometry, trigonometry, calculating devices, algebra, the calculus. 362 illustrations. 1,355pp. 5⅜ x 8½. Two-vol. set. Vol. I: 0-486-20429-4 Vol. II: 0-486-20430-8

A CONCISE HISTORY OF MATHEMATICS, Dirk J. Struik. The best brief history of mathematics. Stresses origins and covers every major figure from ancient Near East to 19th century. 41 illustrations. 195pp. 5⅜ x 8½. 0-486-60255-9

CATALOG OF DOVER BOOKS

Physics

OPTICAL RESONANCE AND TWO-LEVEL ATOMS, L. Allen and J. H. Eberly. Clear, comprehensive introduction to basic principles behind all quantum optical resonance phenomena. 53 illustrations. Preface. Index. 256pp. $5^{3}/_{8}$ x $8^{1}/_{2}$. 0-486-65533-4

QUANTUM THEORY, David Bohm. This advanced undergraduate-level text presents the quantum theory in terms of qualitative and imaginative concepts, followed by specific applications worked out in mathematical detail. Preface. Index. 655pp. $5^{3}/_{8}$ x $8^{1}/_{2}$.
0-486-65969-0

ATOMIC PHYSICS (8th EDITION), Max Born. Nobel laureate's lucid treatment of kinetic theory of gases, elementary particles, nuclear atom, wave-corpuscles, atomic structure and spectral lines, much more. Over 40 appendices, bibliography. 495pp. $5^{3}/_{8}$ x $8^{1}/_{2}$.
0-486-65984-4

A SOPHISTICATE'S PRIMER OF RELATIVITY, P. W. Bridgman. Geared toward readers already acquainted with special relativity, this book transcends the view of theory as a working tool to answer natural questions: What is a frame of reference? What is a "law of nature"? What is the role of the "observer"? Extensive treatment, written in terms accessible to those without a scientific background. 1983 ed. xlviii+172pp. $5^{3}/_{8}$ x $8^{1}/_{2}$.
0-486-42549-5

AN INTRODUCTION TO HAMILTONIAN OPTICS, H. A. Buchdahl. Detailed account of the Hamiltonian treatment of aberration theory in geometrical optics. Many classes of optical systems defined in terms of the symmetries they possess. Problems with detailed solutions. 1970 edition. xv + 360pp. $5^{3}/_{8}$ x $8^{1}/_{2}$. 0-486-67597-1

PRIMER OF QUANTUM MECHANICS, Marvin Chester. Introductory text examines the classical quantum bead on a track: its state and representations; operator eigenvalues; harmonic oscillator and bound bead in a symmetric force field; and bead in a spherical shell. Other topics include spin, matrices, and the structure of quantum mechanics; the simplest atom; indistinguishable particles; and stationary-state perturbation theory. 1992 ed. xiv+314pp. $6^{1}/_{8}$ x $9^{1}/_{4}$. 0-486-42878-8

LECTURES ON QUANTUM MECHANICS, Paul A. M. Dirac. Four concise, brilliant lectures on mathematical methods in quantum mechanics from Nobel Prize-winning quantum pioneer build on idea of visualizing quantum theory through the use of classical mechanics. 96pp. $5^{3}/_{8}$ x $8^{1}/_{2}$. 0-486-41713-1

THIRTY YEARS THAT SHOOK PHYSICS: THE STORY OF QUANTUM THEORY, George Gamow. Lucid, accessible introduction to influential theory of energy and matter. Careful explanations of Dirac's anti-particles, Bohr's model of the atom, much more. 12 plates. Numerous drawings. 240pp. $5^{3}/_{8}$ x $8^{1}/_{2}$. 0-486-24895-X

ELECTRONIC STRUCTURE AND THE PROPERTIES OF SOLIDS: THE PHYSICS OF THE CHEMICAL BOND, Walter A. Harrison. Innovative text offers basic understanding of the electronic structure of covalent and ionic solids, simple metals, transition metals and their compounds. Problems. 1980 edition. 582pp. $6^{1}/_{8}$ x $9^{1}/_{4}$.
0-486-66021-4

CATALOG OF DOVER BOOKS

HYDRODYNAMIC AND HYDROMAGNETIC STABILITY, S. Chandrasekhar. Lucid examination of the Rayleigh-Benard problem; clear coverage of the theory of instabilities causing convection. 704pp. 5⅜ x 8¼. 0-486-64071-X

INVESTIGATIONS ON THE THEORY OF THE BROWNIAN MOVEMENT, Albert Einstein. Five papers (1905–8) investigating dynamics of Brownian motion and evolving elementary theory. Notes by R. Fürth. 122pp. 5⅜ x 8½. 0-486-60304-0

THE PHYSICS OF WAVES, William C. Elmore and Mark A. Heald. Unique overview of classical wave theory. Acoustics, optics, electromagnetic radiation, more. Ideal as classroom text or for self-study. Problems. 477pp. 5⅜ x 8½. 0-486-64926-1

GRAVITY, George Gamow. Distinguished physicist and teacher takes reader-friendly look at three scientists whose work unlocked many of the mysteries behind the laws of physics: Galileo, Newton, and Einstein. Most of the book focuses on Newton's ideas, with a concluding chapter on post-Einsteinian speculations concerning the relationship between gravity and other physical phenomena. 160pp. 5⅜ x 8½. 0-486-42563-0

PHYSICAL PRINCIPLES OF THE QUANTUM THEORY, Werner Heisenberg. Nobel Laureate discusses quantum theory, uncertainty, wave mechanics, work of Dirac, Schroedinger, Compton, Wilson, Einstein, etc. 184pp. 5⅜ x 8½. 0-486-60113-7

ATOMIC SPECTRA AND ATOMIC STRUCTURE, Gerhard Herzberg. One of best introductions; especially for specialist in other fields. Treatment is physical rather than mathematical. 80 illustrations. 257pp. 5⅜ x 8½. 0-486-60115-3

AN INTRODUCTION TO STATISTICAL THERMODYNAMICS, Terrell L. Hill. Excellent basic text offers wide-ranging coverage of quantum statistical mechanics, systems of interacting molecules, quantum statistics, more. 523pp. 5⅜ x 8½. 0-486-65242-4

THEORETICAL PHYSICS, Georg Joos, with Ira M. Freeman. Classic overview covers essential math, mechanics, electromagnetic theory, thermodynamics, quantum mechanics, nuclear physics, other topics. First paperback edition. xxiii + 885pp. 5⅜ x 8½.
0-486-65227-0

PROBLEMS AND SOLUTIONS IN QUANTUM CHEMISTRY AND PHYSICS, Charles S. Johnson, Jr. and Lee G. Pedersen. Unusually varied problems, detailed solutions in coverage of quantum mechanics, wave mechanics, angular momentum, molecular spectroscopy, more. 280 problems plus 139 supplementary exercises. 430pp. 6½ x 9¼.
0-486-65236-X

THEORETICAL SOLID STATE PHYSICS, Vol. 1: Perfect Lattices in Equilibrium; Vol. II: Non-Equilibrium and Disorder, William Jones and Norman H. March. Monumental reference work covers fundamental theory of equilibrium properties of perfect crystalline solids, non-equilibrium properties, defects and disordered systems. Appendices. Problems. Preface. Diagrams. Index. Bibliography. Total of 1,301pp. 5⅜ x 8½. Two volumes. Vol. I: 0-486-65015-4 Vol. II: 0-486-65016-2

WHAT IS RELATIVITY? L. D. Landau and G. B. Rumer. Written by a Nobel Prize physicist and his distinguished colleague, this compelling book explains the special theory of relativity to readers with no scientific background, using such familiar objects as trains, rulers, and clocks. 1960 ed. vi+72pp. 5⅜ x 8½. 0-486-42806-0

CATALOG OF DOVER BOOKS

A TREATISE ON ELECTRICITY AND MAGNETISM, James Clerk Maxwell. Important foundation work of modern physics. Brings to final form Maxwell's theory of electromagnetism and rigorously derives his general equations of field theory. 1,084pp. 5³/₈ x 8¹/₂. Two-vol. set. Vol. I: 0-486-60636-8 Vol. II: 0-486-60637-6

MATHEMATICS FOR PHYSICISTS, Philippe Dennery and Andre Krzywicki. Superb text provides math needed to understand today's more advanced topics in physics and engineering. Theory of functions of a complex variable, linear vector spaces, much more. Problems. 1967 edition. 400pp. 6¹/₂ x 9¹/₄. 0-486-69193-4

INTRODUCTION TO QUANTUM MECHANICS WITH APPLICATIONS TO CHEMISTRY, Linus Pauling & E. Bright Wilson, Jr. Classic undergraduate text by Nobel Prize winner applies quantum mechanics to chemical and physical problems. Numerous tables and figures enhance the text. Chapter bibliographies. Appendices. Index. 468pp. 5³/₈ x 8¹/₂. 0-486-64871-0

METHODS OF THERMODYNAMICS, Howard Reiss. Outstanding text focuses on physical technique of thermodynamics, typical problem areas of understanding, and significance and use of thermodynamic potential. 1965 edition. 238pp. 5³/₈ x 8¹/₂.
0-486-69445-3

THE ELECTROMAGNETIC FIELD, Albert Shadowitz. Comprehensive under- graduate text covers basics of electric and magnetic fields, builds up to electromagnetic theory. Also related topics, including relativity. Over 900 problems. 768pp. 5⅝ x 8¼.
0-486-65660-8

GREAT EXPERIMENTS IN PHYSICS: FIRSTHAND ACCOUNTS FROM GALILEO TO EINSTEIN, Morris H. Shamos (ed.). 25 crucial discoveries: Newton's laws of motion, Chadwick's study of the neutron, Hertz on electromagnetic waves, more. Original accounts clearly annotated. 370pp. 5³/₈ x 8¹/₂. 0-486-25346-5

EINSTEIN'S LEGACY, Julian Schwinger. A Nobel Laureate relates fascinating story of Einstein and development of relativity theory in well-illustrated, nontechnical volume. Subjects include meaning of time, paradoxes of space travel, gravity and its effect on light, non-Euclidean geometry and curving of space-time, impact of radio astronomy and space-age discoveries, and more. 189 b/w illustrations. xiv+250pp. 8⅜ x 9¼. 0-486-41974-6

THE VARIATIONAL PRINCIPLES OF MECHANICS, Cornelius Lanczos. Philosophic, less formalistic approach to analytical mechanics offers model of clear, scholarly exposition at graduate level with coverage of basics, calculus of variations, principle of virtual work, equations of motion, more. 418pp. 5⅝ x 8½. 0-486-65067-7

Paperbound unless otherwise indicated. Available at your book dealer, online at www.doverpublications.com, or by writing to Dept. GI, Dover Publications, Inc., 31 East 2nd Street, Mineola, NY 11501. For current price information or for free catalogues (please indicate field of interest), write to Dover Publications or log on to www.doverpublications.com and see every Dover book in print. Dover publishes more than 400 books each year on science, elementary and advanced mathematics, biology, music, art, literary history, social sciences, and other areas.